Patrick Moore's Practical Astronomy

Other Titles in this Series

Pattern Asterisms: A New Way to Chart the Stars

John A. Chiravalle

With 226 Figures

 Springer

John A. Chiravalle
Safford, Arizona
USA
auggie@aznex.net

British Library Cataloguing in Publication Data
A catalogue record for this book is available from the British Library

Library of Congress Control Number: 2006921158

Patrick Moore's Practical Astronomy Series ISSN 1617-7185
ISBN-10: 1-84628-327-2
ISBN-13: 978-1-84628-327-7

Printed on acid-free paper
© Springer-Verlag London Limited 2006

Printed in the United States of America (TB/EB)

9 8 7 6 5 4 3 2 1

Springer Science+Business Media
springer.com

Acknowledgment

The contents of this catalog were completely photographed and compiled by the author. The mention of the Sky Atlas 2000 by Wil Tirion and Roger W. Sinnott was used in a suggestive manner only. There are many star atlases available that can be used as a guide, but I have found the Sky Atlas 2000 to be most noteworthy. I'm sure many observational astronomers have at least one of these atlases in their possession. All magnitudes mentioned in this catalog are estimates taken from the photographs in this catalog. Because of the large format stellar patterns used, all coordinates in this catalog are approximate, but sufficient to find each object in the sky or planetarium programs on your computer. The simple constellation star maps were generated by the author to give an approximate location of each pattern mentioned. The Mega Star 5 program is one of many excellent computer planetariums available to identify pattern asterisms with. The author referred to this particular program as a proprietor. If you also own a copy, the comments mentioned in this catalog will be of assistance in locating the variable stars, double stars and many pattern asterism shapes before viewing them in your optics.

Much time and consideration was given in choosing each pattern asterism, pattern open cluster and cluster mix to resemble the shape mentioned. Most of these catalog patterns were researched using individuals at star parties, during school tours and visitor/guest presentations at Discovery Park, a science facility located in Safford, Arizona. I would like to thank all of the young people of the Boy's and Girl's Club in Safford, Arizona, for their participation and Peggy McNew and Kay Bailey for their assistance. Ideas and suggestions during this research were used in this catalog to assist the observer. This catalog was a research effort to recognize star groups as a viable astronomical entity. This catalog was also designed to further introduce the use of binoculars into astronomy and also create a new kind of interest during star parties.

Astronomy should be enjoyable as well as educational. I offer to my wife, Dolores, a sincere thank you for enduring over two years of pattern asterisms. I would like to offer praise to a few members of the Desert Skygazers Astronomy Club who rallied around my imagination and forgive others whose vision was insignificant compared to mine and found it necessary to rebuke my findings. I would like to thank the thousands of participants who endured my pattern asterism presentations at Discovery Park and honor those wonderful individuals everywhere who unknowingly contributed their vivid imaginations into this catalog. I do hope that the people who go out at night to seek these shapes in their optics will receive the same joy and excitement I have experienced developing this catalog.

Thank You,
John A. Chiravalle,
Safford, Arizona

Table of Contents

Table of Contents

List of Constellations and Asterisms

	R.A.	Dec.	Size
Cetus			
Hercules Keystone in Cetus	00 hr. 23 min.	−23 deg. 30 min.	$1\frac{3}{4}$ deg.
Pouring Cup	02 hr. 12 min.	+08 deg. 30 min.	$1\frac{1}{2}$ deg.
Cassiopeia			
E.T./Owl Cluster	01 hr. 19 min.	+58 deg. 20 min.	$-\frac{1}{2}$ deg.
Wagon Wheel	01 hr. 45 min.	+61 deg. 20 min.	$1\frac{1}{2}$ deg.
Palm Sander	02 hr. 05 min.	+65 deg. 00 min.	$1\frac{1}{2}$ deg.
Aladdin's Lamp	02 hr. 36 min.	+67 deg. 30 min.	$1\frac{3}{4}$ deg.
Jelly Fish	02 hr. 15 min.	+59 deg. 30 min.	$1\frac{1}{2}$ deg.
Measuring Scoop	03 hr. 27 min.	+71 deg. 50 min.	$1\frac{1}{4}$ deg.
Party Balloon	23 hr. 21 min.	+62 deg. 20 min.	$\frac{3}{4}$ deg.
Bowl of Stars	23 hr. 55 min.	+56 deg. 30 min.	2 deg.
Pisces			
Spatula	01 hr. 35 min.	+08 deg. 00 min.	$1\frac{1}{2}$ deg.
Andromeda			
Pot and Pitcher	01 hr. 55 min.	+41 deg. 20 min.	$2\frac{1}{4}$ deg.
Ursa Minor			
Engagement Ring	02 hr. 32 min.	+89 deg. 16 min.	$1\frac{3}{4}$ deg.
Soaring Owl	16 hr. 00 min.	+74 deg. 00 min.	$2\frac{1}{4}$ deg.
Shark	16 hr. 45 min.	+77 deg. 50 min.	$1\frac{1}{2}$ deg.
Perseus			
Rabbit	02 hr. 37 min.	+55 deg. 45 min.	$\frac{1}{2}$ deg.
Star Eyed Susan	02 hr. 42 min.	+42 deg. 30 min.	$\frac{3}{4}$ deg.
Eridanus			
Letter "F"	03 hr. 38 min.	−07 deg. 45 min.	1 deg.
Letter "D"	03 hr. 39 min.	−01 deg. 45 min.	$\frac{1}{2}$ deg.

	R.A.	Dec.	Size
Taurus			
Dipper Bowl	03 hr. 47 min.	+24 deg. 18 min.	$1\frac{1}{2}$ deg.
Fish Hook	04 hr. 25 min.	+21 deg. 15 min.	$1\frac{1}{2}$ deg.
The Angel	04 hr. 46 min.	+18 deg. 40 min.	$1\frac{1}{2}$ deg.
Poodle	05 hr. 18 min.	+20 deg. 00 min.	$1\frac{1}{2}$ deg.
Barbecue Fork	05 hr. 43 min.	+21 deg. 20 min.	$1\frac{1}{2}$ deg.
Camelopardalis			
Vulture	04 hr. 08 min.	+62 deg. 20 min.	$1\frac{1}{2}$ deg.
Orion			
Hand Gun	05 hr. 36 min.	+09 deg. 40 min.	2 deg.
Lepus			
Funnel	05 hr. 46 min.	−15 deg. 40 min.	$\frac{1}{2}$ deg.
Monoceros			
Christmas Tree	06 hr. 41 min.	+09 deg. 40 min.	$\frac{1}{2}$ deg.
The Tooth	06 hr. 54 min.	−10 deg. 20 min.	1 deg.
Loch Ness Monster	08 hr. 00 min.	−05 deg. 00 min.	$\frac{3}{4}$ deg.
Puppis			
Emission Comet	07 hr. 53 min.	−26 deg. 22 min.	$\frac{3}{4}$ deg.
Hydra			
The Moth	08 hr. 13 min.	−05 deg. 50 min.	1.0 deg.
Umbrella	08 hr. 40 min.	−12 deg. 30 min.	$1\frac{1}{2}$ deg.
Rocking Horse	08 hr. 58 min.	−16 deg. 30 min.	$1\frac{1}{2}$ deg.
The Crown	09 hr. 04 min.	−04 deg. 05 min.	$1\frac{1}{4}$ deg.
Flower Vase	09 hr. 05 min.	−04 deg. 15 min.	1 deg.
The Heart	09 hr. 43 min.	−13 deg. 50 min.	$\frac{3}{4}$ deg.
Vacuum Cleaner	10 hr. 48 min.	−15 deg. 10 min.	$1\frac{1}{4}$ deg.
Sea Horse	13 hr. 10 min.	−23 deg. 30 min.	$1\frac{3}{4}$ deg.
Cancer			
Fish Hooked	08 hr. 28 min.	+07 deg. 20 min.	$1\frac{1}{4}$ deg.
Closed Flower	08 hr. 41 min.	+19 deg. 30 min.	$1\frac{1}{2}$ deg.
Lynx			
The Turtle	08 hr. 54 min.	+35 deg. 00 min.	$1\frac{1}{4}$ deg.
Outline Face	09 hr. 05 min.	+38 deg. 20 min.	$1\frac{1}{2}$ deg.
Ursa Major			
Spade	09 hr. 42 min.	+53 deg. 20 min.	1 deg.
Gas Pump Handle	13 hr. 38 min.	+52 deg. 50 min.	$\frac{3}{4}$ deg.

	R.A.	Dec.	Size
Sextans			
Enter Arrow	10 hr. 50 min.	−09 deg. 00 min.	$\frac{3}{4}$ deg.
Virgo			
Ty's Oil Can	12 hr. 39 min.	−11 deg. 25 min.	$\frac{1}{2}$ deg.
Phaser Gun	13 hr. 20 min.	+03 deg. 00 min.	$1\frac{1}{4}$ deg.
Mortar Trowel	13 hr. 24 min.	− 04 deg. 50 min.	1 deg.
Coma Berenices			
Curved Arrow	12 hr. 53 min.	+19 deg. 30 min.	$\frac{3}{4}$ deg.
Sand Shovel	13 hr. 15 min.	+19 deg. 00 min.	$1\frac{1}{4}$ deg.
Scorpius			
False Comet	16 hr. 53 min.	−41 deg. 30 min.	$1\frac{1}{2}$ deg.
Serpent's Tail	16 hr. 55 min.	−31 deg. 00 min.	$1\frac{1}{4}$ deg.
Butterfly	17 hr. 40 min.	−32 deg. 13 min.	$\frac{1}{2}$ deg.
Draco			
Dog and Stick	17 hr. 36 min.	+68 deg. 40 min.	$1\frac{3}{4}$ deg.
Hercules			
Chinese Kite	17 hr. 58 min.	+45 deg. 50 min.	$1\frac{1}{2}$ deg.
Candle and Holder	18 hr. 03 min.	+48 deg. 00 min,	$1\frac{1}{4}$ deg.
Drill Gun	18 hr. 26 min.	+26 deg. 10 min.	$\frac{3}{4}$ deg.
Ophiuchus			
Kermit the Tadpole	18 hr. 27 min.	+06 deg. 30 min.	$\frac{3}{4}$ deg.
Vulpecula			
Coathanger	19 hr. 26 min.	+20 deg. 00 min.	$1\frac{1}{2}$ deg.
Aquila			
Tennis Racket	19 hr. 50 min.	−04 deg. 35 min.	$\frac{3}{4}$ deg.
Cygnus			
Bent Fan	20 hr. 14 min.	+36 deg. 20 min.	$\frac{1}{2}$ deg.
Horse's Head	21 hr. 32 min.	+48 deg. 30 min.	$\frac{1}{2}$ deg.
Pegasus			
Robin Hood Hat	21 hr. 36 min.	+15 deg. 30 min.	$1\frac{1}{2}$ deg.
The Seedling	21 hr. 43 min.	+10 deg. 20 min.	$\frac{3}{4}$ deg.
Cepheus			
Tobacco Pipe	22 hr. 30 min.	+70 deg. 30 min.	$1\frac{1}{2}$ deg.
The Yacht	23 hr. 25 min.	+64 deg. 15 min.	2 deg.

Introduction to Pattern Asterisms

In the beginning there were stars, lots of them. They filled the sky above the Earth from horizon to horizon, from season to season. These stars were unnamed. They did not possess any recognizable groups or shapes. There were no interesting stories about these bright points of light in the sky until humankind looked up with imagination! Way back then, radios, televisions, cell phones, computers, traffic jams and a busy lifestyle didn't exist to cloud a vision when looking at star patterns. The ancients were quick to envision patterns in bright groups of stars. Their imagination developed gods of many types and powerful wild animals. Some they worshiped and others they feared. Bright objects moved among these star fields in the sky at night and they were named after gods of beauty and war. Unknown to the early humans, these objects are some of our solar system's planets. Guest stars would appear and then slowly disappear. These bright stars were showing their last burst of glory in the form of novas and supernovas. Other objects would appear and develop fiery tails, causing fear among the people. These starry invaders had the ability to change the course of history because of their appearance in the night. Optical aid was not necessary back then to realize that the stellar night was alive with activity. Stars were watched constantly for clues about the future and decisions about the present. That kind of stellar imagination rarely exists today.

The imagination is a wonderful tool that children utilize effortlessly in their daily play. They can use a block of wood to represent a sports car, jet fighter or spaceship. Much of this imagination is put to rest, generally with age. With a little help, this dormant imagination can be revitalized. Images and shapes are everywhere in the stellar sky, just as our ancestor's noticed. This catalog goes beyond what was seen with the unaided eye by those people that studied the sky and were called astrologers. All that's needed is optical aid and some solid coaching to imagine patterns in the night sky. It is time to explore star patterns that was beyond visual reach of those who created the constellations and asterisms. It is time to let your imagination guide you from one star pattern to the next. Set aside small field Messier, NGC and IC objects for a brief time and go wide field! There are star groups in the sky that form wondrous patterns in the eyepiece and they will just take you by surprise. Some are bright and easy in 50 millimeter binoculars, while others are faint and require a 4 inch wide field refractor at the lowest power. These pattern asterisms are represented in all kinds of shapes and sizes a person can imagine. There are dogs, fish, cups, dippers and a hand gun for Orion the hunter. From a small funnel to a soaring bird to an outline smiling face. They appear in all sizes from one-half degree to almost $2\frac{1}{2}$ degrees. It is time to journey into the enjoyable and imaginative sector of astronomy. Looking into the starry sky with binoculars, seated comfortably in a lounge chair and searching for interesting star groups just has to be the most relaxing part of the evening. Just flip through some of these pages and you will see what patterns await your discovery in the sky above.

Just who is capable of seeing patterns in groups of stars? Many early civilizations have interpreted constellations differently. Stellar shapes have been passed on by word of mouth from generation to generation. As wars mingled isolated and geologically divided cultures, their sky lore blended and evolved. They saw the constellations as heroes and warriors equipped with armaments of the day. They carried long spears, pointy clubs and swords into battle as they successfully fought against terrifying monsters and sea creatures. If the constellations were envisioned today, the sky would be filled with high-tech weaponry and super fast flying machines. Indeed, our experiences as individuals growing up may determine how clearly we see a group of stars, or not. Star patterns and stargazing do go together, hand in hand, with a measure of imagination. This creativity has been routed deeply in our past. Professional stargazing may have been set aside as astrology blossomed into astronomy, but we all possess the gift of imagination to some degree. Indeed, the very fields of endeavor we practice in our daily lives may depend upon our imagination, or not. The very toys we enjoyed receiving during special occasions in our youth may have developed our livelihood visions of our future. The kind of patterns you envision seeing in a group of stars may depend upon a variety of complicated factors in your life, but you will see patterns.

An asterism is a group of stars within a constellation. This group may resemble the name of that constellation or just an object that the constellation is identified with. Some good examples are Scorpius the scorpion, the Sickle in Leo the Lion or the Northern Cross in Cygnus. These stars are not related, for the most part, and are just in our line of sight from Earth. Constellations were envisioned thousands of years ago to represent people, animals, sea life and objects of usefulness from those days. Stories of heroism, betrayal, sacrifice, power and knowledge fill the heavens with sky lore. These constellations represented power and ruled the skies with much authority. The astrologer's who were well versed in stellar positions and planetary movement translated these powers for their rulers. Astrologers held prominent positions in the courts of kings and emperors. Much of the imagination that went into naming the stars and constellations a long time ago are used today in our star maps. Astrology may have evolved into the science of astronomy, but there is still much to imagine when looking at star groups. There are endless shapes from the brightest to the faintest stars. From the largest binocular objects to the smallest high powered telescope objects. They are out there waiting to be found and cataloged. This catalog covers pattern asterisms from one-half degree to $2\frac{1}{2}$ degrees. The author is presently involved in a pattern asterism search from 30 minutes to less than 5 minutes in size. These objects will be for telescopes and includes stars to 13 magnitudes. The list will also provide star groups of particular interest designed to spur individual imaginations. Indeed, just when nothing appears left to be discovered and cataloged, an entire stellar world awaits.

Nowadays, all of the constellations, deep sky objects, all that there is to see, are set in print. The novice astronomer excited about his or her new hobby faces a world of computer powered programs not available to a generation of sky watchers before them. Years ago astronomy was all about learning the constellations, finding material and guidance required to lead the way into amateur astronomy. What to look at? Where to look? What size telescope? What power is required to see? All of these questions were a matter of trial and error a generation ago. Astronomy is a completely different hobby today. Robotic telescopes control the night sky during star parties. Computer controlled programming assumes the responsibility for imaging and data taking while the hobbyist sleeps. Beginning amateur astronomers only need to know some of the brightest star names and their general location to guide the electronic data-based telescopes to thousands of objects in the sky. The hobby of astronomy has futuristically leaped forward from imagination to computerization.

Pattern Asterisms

If you have a pair of binoculars, and many people do, these asterisms are visible for your enjoyment and education. Binoculars are excellent first telescopes because they are wide field, easy to steady and most importantly, you use both eyes. That means your brain receives more information and perceives faint stars more clearly then using one eye. A good pair of 50 millimeter binoculars that vary in power is a good observational telescope. Attach a small red dot finder on your binoculars, a steady mount, a quality star atlas and navigating the stars is a pleasure using the simple information in this catalog. If you do have a wide field refracting telescope that covers a 2 degrees or more field of view, and there are many 4–6 inch (102–152 mm) scopes of F6 to F7 that will fit this category, you are in luck! All of these asterisms can be viewed with these telescopes.

Binoculars used to view the stars in the night sky must be held very steady. Leaning against the side of a wall or pole will only provide temporary stability when viewing. A mount of some kind will be needed to hold the scope constant. Another problem inherent when using binoculars to view straight up in the night sky is the uncomfortable head and neck position. Some binoculars overcome this situation by providing a 45 or 90 degree diagonal eyepiece setup. For those binoculars not equipped with angled eyepieces, an adjustable lounge chair will prevent neck strain and increase stellar pleasure. Never the less, a simple support will be necessary most of the time. The most basic and inexpensive mechanism for maintaining solid binocular positioning is as close as your broom closet. A long handled staff or stick (Fig. 1.1) is all that is required for personal viewing. By angling the staff, an individual can adjust the binocular viewing height. One hand supports the staff and steadies the pair of binoculars resting on that hand. The second hand holds and adjusts the direction of the binoculars. The staff maintains a steady base while the binoculars reside between the staff.

Figure 1.1. Johnny Lopez being shown with a binocular staff support.

The staff support is feasible for viewing from the horizon to about 45 degrees skyward. With some minor woodworking skills, an angle staff mount (Fig. 1.2) can be assembled. This will increase viewing angle without increasing the effort. It is composed of two sections of staff with an angle connection between them. The angle staff mount can be used by small and tall individuals by simply moving up the staff.

Figure 1.2. Tyler Scott Farrell demonstrating an angle binocular staff support.

Figure 1.3. Corie Johnson and Keoin Gonzales demonstrating a binocular parallelogram mount.

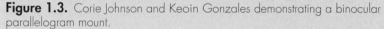

If there is a wall to lean against, this too will increase binocular stability. These two mounts are simple and practical for individual use.

For group viewing, an ingenious binocular mount can be found almost anywhere stellar optics is sold. The "parallelogram" mount (Fig. 1.3) allows the binoculars to be moved in a vertical position to suit the individual's height. This vertical movement of the mount will not effect the position of the object being viewed. They are convenient for both standing and lounge chair positions. Like most particular contraptions, they can be expensive, but most impressive during star parties. There are different size parallelogram mounts to handle medium to large binoculars. Scanning the skies with a large pair of binoculars is like looking through the porthole of a spaceship during a star cruise. With the proper mount attached to these twin scopes, the stellar voyage will no doubt be more pleasurable.

To judge stellar perspectives, I used 50, 80 and 100 millimeter binoculars. For this catalog, I used a 5 inch (127 mm) refractor at F/6.5 and a 2 inch (51 mm) eyepiece to search for star groups. This setup gave me almost a 3 degree field of view at 16 powers. A wide field of view and low power is the key to unlock isolated star groups. To photograph the sky I used a 4 inch (102 mm) F/6.5 refracting telescope at prime focus on the same mounting and switching eyepieces on my 5 inch (127 mm) refractor to use as a guide telescope. I exclusively used Kodak Professional T400 CN black and white film to photograph the star images. After developing, the negatives were placed on disk. I placed these photos in the Adobe Photo Shop program to erase

Figure 1.4. Groups of stars in our line of sight.

unwanted space debris, satellites and aircraft from the exposures. I also revitalized the pictures and added informational data to them. Some exposures were only two to four minutes, depending upon background star interference. Some images that are viewed in binoculars and small telescopes as clearly defined patterns can easily be washed out with timed exposure photography. One good example is the Rabbit asterisms in Perseus, Fig. 2.7. In my 16 inch (416 mm) Newtonian at 32 power the Rabbit shape is obvious. A six minute exposure of the Rabbit through my 4 inch (102 mm) refracting telescope completely washes out this pattern. The Rabbit happens to be part of the open cluster TR-2. Care and many extra exposures were taken to ensure that the photographic image resembles what is seen in the eyepiece. Despite this effort, faint background stars do appear on photographs. In most cases, these faint background stars seen on film will not interfere with the asterism's shape or be seen when viewing with binoculars and small telescopes.

So, what in the world is a pattern asterism and are they really new? Of course not and you see them all the time in the sky when viewing. A few asterisms are commonly known while most in this catalog are brand new. Visually, the Pleiades open star cluster, M 45, is a fine example of a pattern, Fig. 3.2. Not commonly known is the bright open

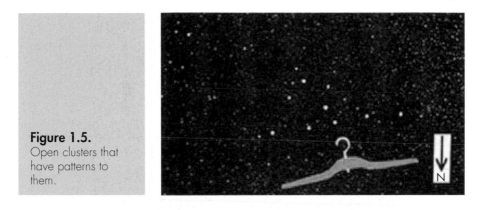

Figure 1.5.
Open clusters that have patterns to them.

star cluster in Orion called CR-69, Fig. 3.8. With some dot to dot lines and clip art representing the pattern shape, a hand gun becomes completely visible in this group of stars. For a presentation involving third graders, I call CR-69 a water gun. It is just a matter of using the correct ammo to suit the age group.

As a hazy spot in the night sky, M 44 in Cancer becomes a closed flower, Fig. 5.5 with low powered binoculars. To provide a mixture of enjoyment with education, I have incorporated humor with most asterisms. It is my philosophy that both astronomical enjoyment and education go hand in hand for a memorable experience. Most of these pattern asterisms in this catalog will be new to everyone and a few will be old hat. Some patterns are actually open clusters that have been viewed by every astronomy hobbyist, but they may not have connected the dots to visualize an object. In most cases these patterns become more apparent at very low power. With open star clusters it is important to only use the outer stars to shape the pattern.

The pattern asterisms in this catalog are organized into three main groups. The first group is a random set of stars (Fig. 1.4) of varied magnitudes, but all of these stars are in our line of sight. This group is generally bright, stands out by itself and is not confused by any faint background stars around it. The stars in the pattern may also be faint, but the stars dot the area in alignment which causes the pattern to stand out in the field. Then there are the imaginative few that require good principles of observing; a test to challenge the best. The second group consists of well known open clusters, some large and some small (Fig. 1.5) but most are one-half degree or larger. There are a few smaller exceptions that I absolutely refuse to leave out of this catalog because of their popularity. The third group combines the best of the best: A pattern asterism open cluster mix that combines unrelated stars to gorgeous open clusters to form an object or relates the two in a theme (Fig. 1.6). In one case, the Sand Shovel pattern asterism, Fig. 6.6, in Coma Berenices has a bright globular cluster, M 53, in the same field. The shovel is said to shower the Berenices cluster with star dust. The Sand Shovel is visible in 10 powers with 50 millimeter binoculars. Small to medium telescopes will do justice to M 53 by revealing the faint outer stars. In another example, an emission nebula NGC 2467, substitutes for the head of a comet, Fig. 4.4, in Puppis.

I have no doubt some of these pattern asterisms will gravitate to your liking. A few will become your favorites and one or two will simply knock you socks off! These

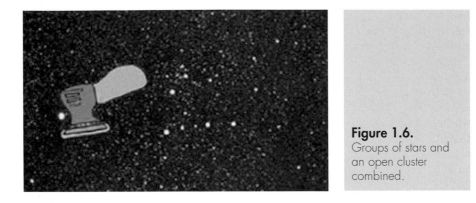

Figure 1.6.
Groups of stars and an open cluster combined.

patterns have been laced with sci-fi humor because most stellar thinkers are also futuristic minded. They are accessible to beginning amateur astronomers with very little optical aid. These asterisms also offer a welcomed break in routine for those serious observational astronomers. During star parties these objects will act as the icing on the stellar cake. In any event, if you are an old sky astronomer like I am, you will have some viewing pleasures awaiting you in this catalog. If you're new to the hobby of astronomy, this may be the kicker to get you fully involved. Some of these asterisms will guide you right to some pleasant deep sky objects, starting off your hobby with many surprises. The optical aid to view most of this catalog is already in most people's homes in the form of closet kept binoculars.

Steering the Course

You certainly would not travel to a distant, unfamiliar town by car without a road map. Getting lost on the road is all too easily accomplished. The same is true for traveling among the stars. As mentioned previously, go-to, computerized telescope make finding objects so much simpler when properly aligned. They can also be used to locate these asterisms. The coordinates given for each pattern will take your properly adjusted telescope right to them. Telescope mounts are becoming more versatile every year. In the near future some go-to telescope mounts will operate by voice commands! Until then, setting circles, knowledge of the constellations and a good star atlas on paper or software for your computer will do. In fact, a good planetarium program on your computer will also show you these asterisms. I used the Mega Star 5 planetarium program to see how the patterns I observed in the telescope appeared on my computer. I then mentioned this similarity in my notes. Remember, these shapes in the sky are large by telescope standards. Quite a few of these asterisms are visible in telescope finder scopes. This catalog was not designed to test your stellar ability, but for you to enjoy and develop your education into astronomy, a hobby that may be brand new to you. Most beginning amateur astronomers develop frustration searching for objects much too faint for their equipment. Observing these patterns will not require searching for objects fainter then your optics can resolve. Once you become accustomed to these pattern asterisms in this catalog, you may very well develop a few of your own. No matter what system you use to find these asterisms, whether you use the old fashion star hopping or modern go-to computerized methods, be prepared to envision shapes and friendly patterns everywhere in the sky.

Imagination is the key to go from dot to dot among the stars. Some patterns are obvious just as they are named. But beware, star pattern shapes are seen differently by individuals. For instance, the Robin Hood Hat, Fig. 8.1, may also be seen as a Cowboy Boot and that is where the enjoyment begins. Some extra benefits will unfold when searching for pattern asterisms. They are mixed with faint deep sky objects that the beginning amateur astronomer will find automatically during the asterism search. Suddenly, finding a deep sky object, in some cases, will be as easy as finding an asterism. For example, Ty's Oil Can uses M 104, the Sombrero Galaxy, Fig. 6.2, as the oil spill. Find the asterism Ty's Oil Can in your telescope and right next to this pattern is a bright galaxy. There are multitudes of beautiful open clusters that can easily be found in this manner. Along with the deep sky objects, an entirely new quest to guess the asterism's shape can be offered. Just how many different images can be seen from one group of stars? During a star party, after everybody has had their look in the eyepiece and taken a guess, the given catalog name and pattern chart from this book can be revealed. Watch their expressions of surprise as they check the eyepiece once again, comparing the traced out pattern and clip art from the catalog. It will add a new twist to the same old star party. People who thought little of using binoculars to look at stars will see astronomy in an entirely new way. They will, in some cases, develop an enthusiasm toward star groups. I've found out that astronomy hobbyists, once exposed to pattern asterisms, will start to see patterns in the stars everywhere.

Parameters

Finding a large group of stars and connecting a series of lines to some of the bright stars in the group to produce a pattern is **not** what this catalog is about. Many groups of stars were discarded from this catalog because they did not represent a defined shape or the star patterns were not isolated enough. The star groups forming patterns must be by themselves or bright enough to stand out from the background stars. They cannot form a partial pattern, but must be reasonably complete. In some cases like the Tooth asterism, Fig. 4.2, in Monoceros, a string of stars outlining this shape will stand out most obviously from the surrounding stars. An open cluster as a pattern is defined by the outer stars only. This rule is valid in both open clusters and asterism/cluster combinations. Although it requires a bucketful of imagination to find these pattern asterisms, once lined on the negative, little imagination will be required to see them in the eyepiece. For more assistance, clip art has been added to most patterns in this catalog. Clip art is important to let the observer **see** the pattern I am visualizing. Pattern size was selected to utilize spotting scopes, binoculars and wide field telescopes. To that end, fields of view from one-half degree to $2\frac{1}{2}$ degrees were elected as a boundary. It is very important to put boundaries to a group of stars in a pattern. The size of the asterism is marked on the data plate to the side of the photograph. The sizes of the asterisms are also marked on the index lists. Please note that this size does not represent the size of the photograph displayed, but the asterism itself.

Umbrella $1\frac{1}{2}$ deg. 8 hr. 40 min. in Hydra in size −12 deg. 30 min.

*Note: Flamsteed numbering system Double stars

1. Provide a negative picture with the positive picture.
2. Line out the image (dot to dot) on the negative.
3. In most cases, include a clip art representation of the pattern asterism.

4. Provide size, direction, magnitude, star count and Sky Atlas 2000 location.
5. Suggest size of optics to see pattern asterism.
6. Curious or humorous details about the pattern asterism.

Note: Flamsteed numbering system, variable stars and double stars used for identification of pattern asterisms.

Figure 1.7. Pattern asterisms made easy.

Another factor to consider was magnitude. There should be bright pattern asterisms less than 9th magnitude, fairly bright pattern asterism mixes (open clusters and unrelated stars in our line of sight) under 10th magnitude and challenging faint pattern asterisms no fainter then 12th magnitude. Most of these patterns contain faint and bright stars to totally outline their shape. In most cases the bright stars can hold their patterns together. A few groups are made up of faint stars strung together to form a pattern. With this range of magnitude, common binoculars can be used to search for these asterisms. Each asterism has a detailed report to assist the observer (Fig. 1.7). As mentioned above, there are three kinds of patterns in this catalog. Groups of stars that form patterns, open clusters that have patterns, usually at low power, and asterism, open cluster mixed together Normally we see stars as a constant brightness in the sky, but that is not often true. There are stars that vary in brightness all the time and this fact must be taken into consideration when we are looking at star patterns. Granted, most stars vary in brightness only a few tenths of a magnitude and that would not overly distort a shape of a pattern. But, it is interesting as well as educational to be aware of this fact. To this end, variable stars are noted in the asterisms that contain them. Also, information is provided about these variable stars and their effects, if any, concerning the pattern asterisms. Included on most asterisms are double stars. Double stars are plotted for positional location when using a planetarium program on your computer and because they may be of observational interest. If there is no double star involved in an asterism, a star not connected to the pattern may be used for reference.

Photographing these groups of stars as they would appear in the eyepiece was of major consideration. Over-exposure would tend to wash out some patterns or cause stellar crowding. Some patterns in the Milky Way were partially sensitive in this manner. All of the patterns are reproduced as they would be seen in binoculars and small telescopes. True north is given on each photo, but the pattern is viewed upright, in most cases, for visual effect and clarity. All magnitudes and pattern sizes are approximately to assist the observer. Flamsteed numbers, when available, are given for stellar location in a particular constellation when using a star catalog. General constellation maps are provided in this catalog to show all of the pattern asterisms in that constellation. The individual observer must take into consideration their light polluted sky conditions and the size of their optics when searching for pattern asterisms. Optical suggestions are recommended for each asterism under dark skies only. A stand or mount is advised to steady binoculars when using 12 magnifications and above. A binocular mount can be as simple as a long stick or pole with the binoculars resting on your hand grasping the pole, as noted above. This catalog is oriented in right ascension except when necessary to group asterisms involved in one constellation. All of the photographs in this catalog were taken by the author.

Catalog Index

This catalog has three main indexes. The *List of Constellations and Asterisms* provides the main layout of this catalog from 00 hours in right ascension, and by constellation. However, the flow is interrupted in a few constellations so that the asterisms can be condensed. This list gives the constellation and the asterisms located in each constellation. It also gives the right ascension and declination of each asterism and the size of each asterism.

There is a *Glossary of Deep Sky Objects* that are involved in asterisms, deep sky objects that form patterns or relating to a particular asterism. Although most of the deep sky objects in some asterisms are open clusters, this is not true for all deep sky objects. There is a galaxy involved in Ty's Oil Can, a globular cluster in the Sand Shovel and an emission nebula as the head of the Emission Comet in Puppis. There is a glossary of double and variable stars also provided.

Asterism at Zenith by Month covers each asterism at its highest point in the sky for Northern observers during the year. This list also provides the reader with optical size and power requirements capable of seeing the asterism in dark skies. Many factors can exist to offset optical size suggestions in this catalog. Humidity, inversions layers, air turbulence, sky smog, dust and the presence of lunar glow can be presence, just to mention a few factors.

Main Body of Catalog

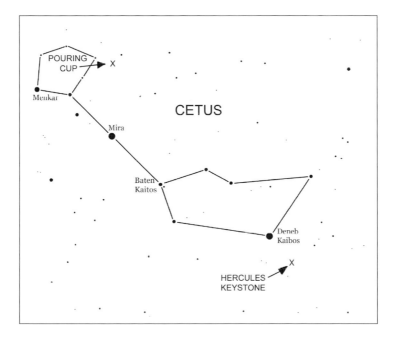

POURING CUP → X

Menkar

CETUS

Mira

Baten Kaitos

Deneb Kaibos

HERCULES KEYSTONE → X

The Hercules Keystone in Cetus

00 hr. 23 min.   −23 deg. 30 min.

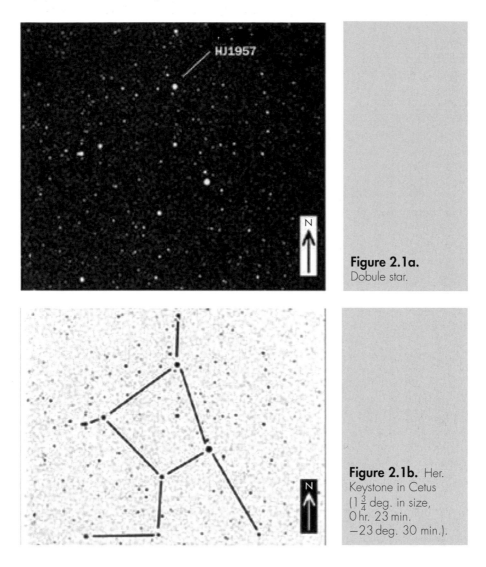

Figure 2.1a.
Dobule star.

Figure 2.1b. Her. Keystone in Cetus (1¾ deg. in size, 0 hr. 23 min. −23 deg. 30 min.).

Here's one for the record books! An asterism of an asterism, Fig. 2.1, and why not; imitation is the best form of flattery. This keystone look-a-like covers a field of view 1¾ degrees and is easily visible in 50 millimeter binoculars. I suggest 10 to 20 powers for this bright asterism. The Keystone pattern stands out nicely with few foreground stars to obstruct its shape. A true Hercules Keystone in the belly of the Whale, figuratively speaking that is. All of these stars are brighter than 10 magnitudes, one star being at 7 magnitude and three stars at 8 magnitudes. This asterism is located 7 degrees southwest of star 16, Deneb Kaitos and the four main stars of the Keystone are visible on the Sky Atlas 2000. On the same atlas, galaxy NGC 45 is only $1\frac{1}{2}$ degrees west of the Cetus Keystone. This is a very faint galaxy only visible in large scopes. On the Mega Star 5 planetarium program, the Cetus Keystone shape appears identical to the photo at 9 magnitudes.

Unfortunately, this asterism is placed well south for most upper Northern observers. However, if you can locate M 4 in Scorpius or M 41 in Canis Major, for example, then the Keystone in Cetus should be visible on a clear, low humidity night, well placed high in the December sky. There is one multi-double star that has a 7.5 magnitude star and a 9 magnitude star separated by 6 seconds of an arc, but that is a telescope object. If you are a true constellation buff, this little Keystone mini-me will be most delightful. No imagination is required to envision this group of stars. Be the first one on your block to surprise your astronomy friends with this carbon copy asterism of the Hercules keystone.

Pouring Cup

02 hr. 12 min.   +08 deg. 30 min.

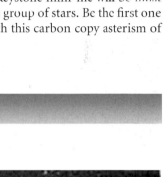

Figure 2.2a.
•Note: Flamsteed numbering system.
°Note: Variable star, Double star.

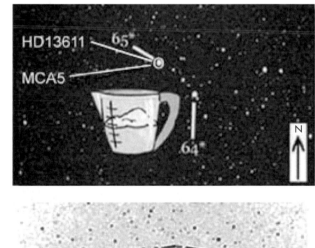

Figure 2.2b.
Pouring Cup in Cetus ($1\frac{1}{2}$ deg. in size, 2 hr. 12 min. +8 deg. 30 min.).

Figure 2.2c.
Pouring Cup
in Cetus.

Show me a kitchen well supplied and I'll point out a pouring cup, Fig. 2.2, to measure ounces and milliliters. At $1\frac{1}{2}$ degrees in size, Cetus the Whale must measure Krill mouthfuls at a time. This asterism is composed of two very bright stars, star 65 Cetus at magnitude 6 and star 64 Cetus at magnitude 7. Both of these stars represent the upper pouring end of the cup. Forming the top of the handle are two stars, magnitude 8 and magnitude 8.5. The remaining stars are not fainter then 10.5 magnitude. There are 14 stars used to trace out this asterism. Four stars are visible on the Sky Atlas 2000. At 10.5 magnitudes, the Pouring Cup is plainly visible on the Mega Star 5 planetarium program. Sixty millimeter binoculars at 16 powers at the steady (on a support) should do the trick.

This asterism will require back and forth glances at the positive photo in this catalog and in your scope. The pouring end of the cup, star 65 is a variable star as well as a tight double star. Since the variable star HD13611 only varies slightly, no change will be noticed in the asterism. The clip art will provide lots of measuring help to see the image. Some asterisms jump into plain view while others need closer inspection. But, like a sudden bright idea mentally appearing, once you see it in your optics, it's yours to show at star parties. Finding this asterism will not turn you into a good cook, but it will develop a fair measure of stellar pleasure.

Cassiopeia Map and Asterisms

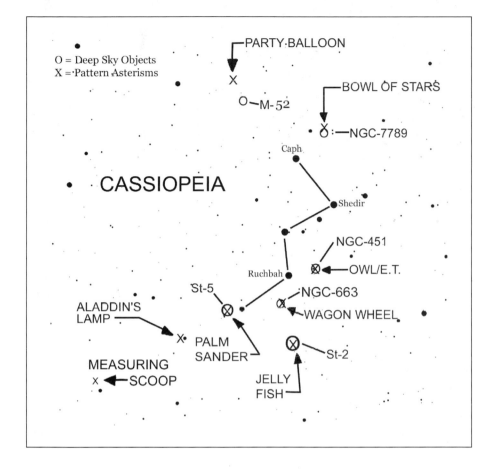

O = Deep Sky Objects
X = Pattern Asterisms

PARTY·BALLOON

X

O—M-52

BOWL OF STARS

☿:—NGC-7789

Caph

CASSIOPEIA

Shedir

NGC-451

Ruchbah

OWL/E.T.

St-5

NGC-663

ALADDIN'S
LAMP

WAGON WHEEL

X· PALM
SANDER

St-2

MEASURING

x ←SCOOP

JELLY
FISH

E.T./Owl Cluster

01 hr. 19 min.   +58 deg. 20 min.

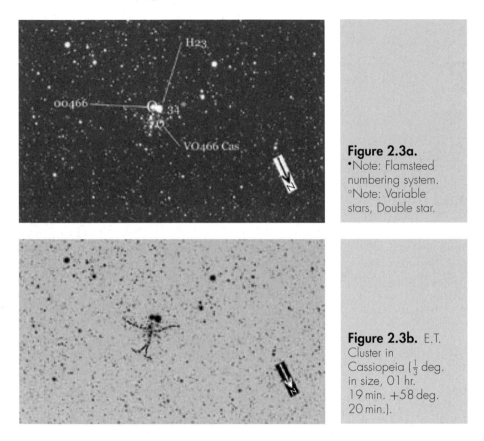

Figure 2.3a.
•Note: Flamsteed numbering system.
°Note: Variable stars, Double star.

Figure 2.3b. E.T. Cluster in Cassiopeia ($\frac{1}{3}$ deg. in size, 01 hr. 19 min. +58 deg. 20 min.).

At first glance, I would say that the movie company put this little guy up in the sky to advertise *The Extraterrestrial* movie way back when. NGC 451 at 20 minutes in size, Fig. 2.3 , is too small to fit my parameters of one-half to $2\frac{1}{2}$ degrees for this asterism catalog. On the other hand, I would not abandon this petite pattern open cluster. After all, we never leave our little aliens behind! It does require more power then the normal binoculars provide. Aperture wise, a good spy binoculars having 60 millimeters and 30 powers on a mount would suffice. Any run of the mill refractor or reflector will show this out-worlder waving its arms frantically needing to call home. As far as needing an imagination, that is really not necessary. During star parties, this little being always steals the movie – I mean the show.

Two bright stars, 5.5 magnitudes and 7.5 magnitudes, make up the Owl's eyes or the big eyes of the Extraterrestrial. The brightest star carries the Flamsteed number 34 on the Sky Atlas 2000. Also, star 34 is a multi-double star (H23) . The other 'eye star' is a variable and you would not notice its faint wink. One other variable star, VO466 Cas, that starts the left arm does wave at almost 1 magnitude in brightness. Besides that distinction, this cluster is invisible in a 50 millimeter finder scope at 10 powers. Star

34 Cassiopeia is the bright eye (of the Extraterrestrial) and a good location device. The shape becomes apparent as a group of 10.5 magnitude stars pop into view with enough power. Once again, most telescope wielding astronomers worth their salt have had this good buddy in their sights many, many times. For the first timers out there, this pattern open cluster should be top priority on their must – see object's list. Don't stop until you have found this little guy in the eyepiece because the search is a rewarding find.

Wagon Wheel

01 hr. 45 min.   +61 deg. 20 min.

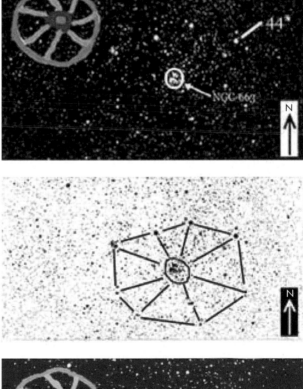

Figure 2.4a.
•Note: Flamsteed numbering system.

Figure 2.4b.
Wagon Wheel in Cassiopeia
($1\frac{1}{2}$ deg. in size,
1 hr. 45 min.
+61 deg.
20 min.).

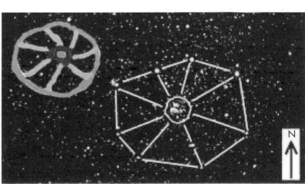

Figure 2.4c.
Wagon Wheel in Cassiopeia.

This is only one of many fine examples of an open cluster, asterism mix. In fact, two open clusters play an important roll, no pun intended, in this pattern. The random stars form an oval around NGC 663 which portrays, in this case, the hub of the Wagon Wheel, Fig. 2.4. NGC 654, another open cluster, appears as the broken or loose spoke in the wheel. This is a bright asterism mix visible in a 50 millimeter finder scope at 10 powers. The Sky Atlas 2000 shows most of the stars clearly on the map. All the stars in the wheel are 8.5 magnitude or brighter. The brightest star in the wheel is labeled star 44 Cassiopeia. Of course, there are plenty of background stars of the Milky Way in this shape, but the pattern stands out well. As noted, a pattern or shape is in the eye of the beholder. I see a wagon wheel and some people see a pizza pie. Okay, maybe it is time for dinner, you never know. Others see an orange sliced in half. Vitamin C is good. How about a grapefruit? A little tart for me, but if you are on a diet, that would bring on the representation, no doubt. For those little multi-legged creatures, this design could be a spider's web.

At $1\frac{1}{2}$ degrees in size, a nice wide field 4 inch (102 mm) refractor with a high millimeter eyepiece will bring out both open clusters well. A nice 4 inch (102 mm) mounted binoculars at 20 or 25 power would do better. The open cluster NGC 663 has been thoroughly investigated and contains many double stars. This area of Cassiopeia simply beats down the door with star groups. Making patterns out of those star groups is another matter. The Wagon Wheel is not difficult to see with a little help from the photo, but this all depends upon your scholastic ability in first grade doing your numbered dot to dots.

Palm Sander

02 hr. 05 min.   +65 deg. 00 min.

This woodworking machine, Fig. 2.5, is powered by a faint open cluster called Stock 5 which happens to be located in the motor housing, not at all an unusual place to find heavy stellar metals. 52 Cassiopeia is the center star of the sanding pad and 53 Cassiopeia is the front of the motor housing. This asterism measures $1\frac{1}{2}$ degrees

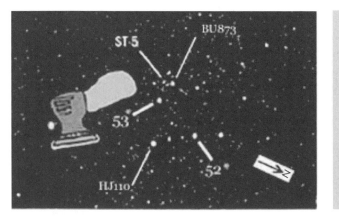

Figure 2.5a.
•Note: Flamsteed numbering system, Double stars.

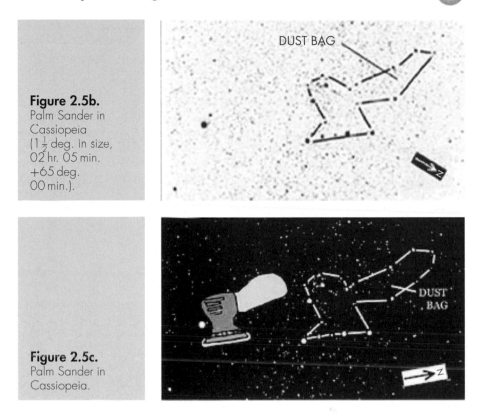

Figure 2.5b.
Palm Sander in Cassiopeia ($1\frac{1}{2}$ deg. in size, 02 hr. 05 min. +65 deg. 00 min.).

Figure 2.5c.
Palm Sander in Cassiopeia.

in size and is considered a smooth catch with all of its bright stars. Four bright stars represent the sanding pad, the dust bag has five stars that stand out from the Milky Way haze and the motor housing has four stars. The shape of this asterism does not require much guesswork, although a chain saw image is also visible using the dust bag as the blade. Of course, a leaf blower might not be too hard to see either. In any event, not too far off the beaten path since each shape does deal in wood removal or tree leftovers. The Palm Sander displays well on the Mega Star 5 planetarium program at 10.5 magnitude.

Finding the Palm Sander is as easy as finding a knot in pinewood. This asterism sits on a starry shelf only 1 degree east of the lazy arm of the W and star 45 Cassiopeia. A double star (BU873) is located just about where the power switch, on some models, would be placed. The double would be for on and off, of course. Fifty millimeter binoculars at 16 powers will display all of the stars. There is another double (HJ110) located on the front sanding pad. It has a 5.2 magnitude star and a 9.5 magnitude star separated by less then a minute of an arc. The faintest star in this group is 9.5 magnitudes. Low F ratio reflectors are suitable for these moderate size asterisms. Using a 4–6 inch (102–152 mm) Newtonian at F/4 to F/6 with a 40 millimeter eyepiece will do justice to this asterism. The blanket of faint stars in the field of view would represent the saw dust from a smoothly sanded view.

Aladdin's Lamp

02 hr. 36 min.   +67 deg. 30 min.

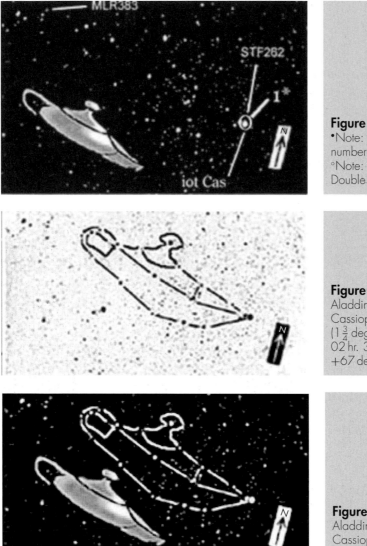

Figure 2.6a.
•Note: Flamsteed
numbering system.
°Note: Variable star,
Double stars.

Figure 2.6b.
Aladdin's Lamp in
Cassiopeia
(1¾ deg. in size,
02 hr. 36 min.
+67 deg. 40 min.).

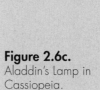

Figure 2.6c.
Aladdin's Lamp in
Cassiopeia.

Be careful what you wish for. You may wish I had never found this asterism because it requires a bucketful of imagination to assimilate. On the other hand, the Aladdin's Lamp is an interesting group of bright stars, easy to find. Simply follow the lazy arm of the W one full arm's length and it will take you to star number one Cassiopeia where the Aladdin's Lamp resides. Star one is the front of the lamp. That star is also a double and variable. Its variability is not noticed in small telescopes. The top star in the handle of the lamp is also a double star. The general shape of the lamp is outlined by 9 magnitude stars and brighter. This $1\frac{3}{4}$ degree asterism is visible in a 50 millimeter finder at 10 powers. Nine stars are visible on the Sky Atlas 2000, but the lamp shape is not apparent at all.

Look carefully at the negative photo to see how the stars are connected and then try to relate that view to the positive photo. You will notice how the front of the lamp almost forms itself with stars and how the back handle is curved in 10.5 magnitude stars. A 7.5 magnitude star represents the dome of the lamp. A 4 inch (102 mm) refractor at low power will bring out all of the features of this asterism. This is a bright asterism of stars and should bring lots of grunts and I-don't-know (s) from the crowd at star parties. Since this asterism has a double star in front and in back, just two rubs on the lamp will bring out the Genie. Perhaps the Genie of the lamp can help show you what you're looking at?

Jelly Fish

02 hr. 15 min.   +59 deg. 30 min.

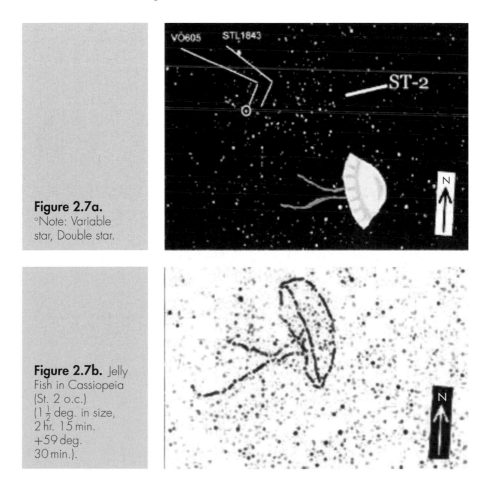

Figure 2.7a.
°Note: Variable star, Double star.

Figure 2.7b. Jelly Fish in Cassiopeia (St. 2 o.c.) ($1\frac{1}{2}$ deg. in size, 2 hr. 15 min. +59 deg. 30 min.).

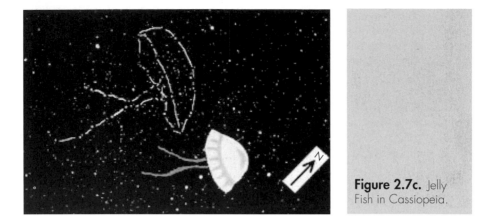

Figure 2.7c. Jelly Fish in Cassiopeia.

Officially known as Stock 2, this pattern open cluster goes by the name of the stick man for most amateur astronomers, and rightly so. I can see the stick legs, body and arms but no head. The headless Halloween horseman open cluster perhaps? Okay, I could see how easily I can lose my head talking about this pattern. This group forms, in my opinion, the Jelly Fish shape, Fig. 2.7, completely at 10 magnitudes. The shape starts at around 8.5 magnitudes. Photographing this cluster of stars to show the Jelly Fish as it appears in binoculars was no easy matter. This cluster is buried in the Milky Way and too long an exposure will wipe out the shape completely. I used different timed exposures, and then chose the film frame most identical to the view in the eyepiece.

The Jelly Fish is $1\frac{1}{2}$ degrees in size and requires a 60 millimeter mounted binoculars at 16 powers. A 3 inch (76 mm) refractor will do nicely. The shapes will just start to appear in a 50 millimeter finder scope at 10 powers. There are many double stars in this cluster and I picked out just one for reference. The variable star VO605 only alters 0.2 magnitudes in brightness. The few other variable stars in this cluster will not alter the shape of the Jelly Fish. The Jelly Fish shape appears clearly on the Mega Star 5 planetarium program at 10 magnitudes. Finding this pattern open cluster is as easy as finding the double cluster in Perseus. There is a bridge of stars that lead from the double cluster right to Stock 2. Therefore, the Jelly Fish managed to swim a little over 2 degrees north, just inside the Cassiopeia boarder. Gee whiz, I hope The Queen is not in the mood for cnidarian!

Measuring Scoop

03 hr. 27 min.   +71 deg. 50 min.

The Measuring Scoop, Fig. 2.8, is a cute pattern asterism at $1\frac{1}{4}$ degrees in size. It pulls no punches and requires absolutely no imagination to see. The scoop is bright with all of its stars no fainter then 8.5 magnitude. It is visible in any closet kept binoculars above 8 powers. Like normal, more power would be nice. It looks like a coffee scoop or a saucepan that experienced a bad case of metal fatigue, or maybe a bad heat day. This asterism is voted as a good choice at any star party for guess and tells. At

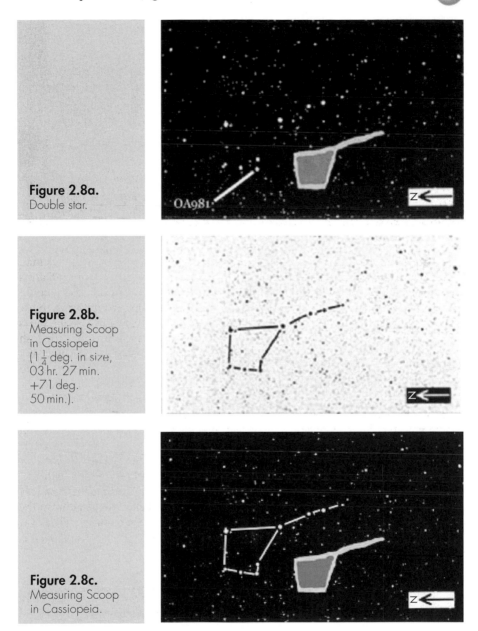

Figure 2.8a.
Double star.

Figure 2.8b.
Measuring Scoop
in Cassiopeia
$(1\frac{1}{4}$ deg. in size,
03 hr. 27 min.
+71 deg.
50 min.).

Figure 2.8c.
Measuring Scoop
in Cassiopeia.

9.5 magnitudes the Measuring Scoop is visible with 11 stars detailing its shape on the Mega Star 5 computer program. The Measuring Scoop is completely visible in the Sky Atlas 2000 with nine stars dotting out the image.

To locate the Measuring Scoop, just star hop from the lazy arm of the W past the Palm Sander and the Aladdin's Lamp. I did not plan it that way, but if that's the scoop, I'll use it. This pattern does not contain any variable stars, but does contain one tight double star located at the bottom of the bowl, suggesting two scoops per strong cup of coffee. There is one other thing I should mention about this asterism and that is it looks like a kite, almost. In the negative photo it looks more like a kite with a long tail then a scoop. It can also look like a meat cleaver, but that's cutting it too close. So, be prepared to hear the word kite with group viewing. Some people will say that it looks like a kite more then a scoop. People will see what they will and that is a good thing. Asterisms belong to all of us in whatever fashion we see them in.

Party Balloon

23 hr. 21 min.   +62 deg. 20 min.

Just how many times have you seen this happen at a birthday party? A balloon filled with helium, tied with some long streamers is accidentally released into the sky. This is the concept envisioned as the Party Balloon, Fig. 2.9, flies passed M 52 in Cassiopeia. M 52 is one beautiful open cluster and it is used as a backdrop for this pattern asterism, making a serene portrait. Lots of bright stars make up the Party Balloon rendering it easy in 50 millimeter binoculars at 16 powers. Because of its three-quarters of a degree size, a multitude of optical hardware can be used to enjoy this vista. At 9.5 magnitudes

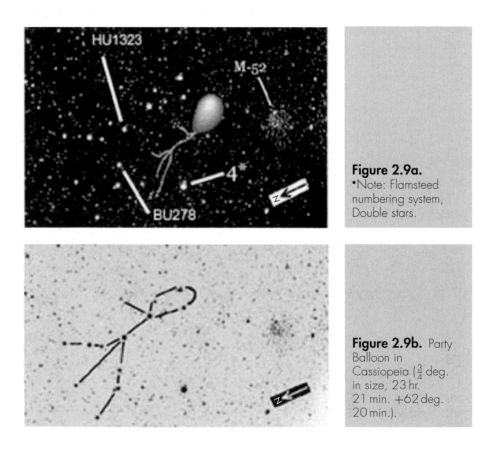

Figure 2.9a.
•Note: Flamsteed numbering system, Double stars.

Figure 2.9b. Party Balloon in Cassiopeia ($\frac{3}{4}$ deg. in size, 23 hr. 21 min. +62 deg. 20 min.).

Figure 2.9c.
Party Balloon in
Cassiopeia.

the shape appears and one magnitude more completes all of the steamers. Most of the shape is visible on the Sky Atlas 2000 right next to star 4 Cassiopeia which is not part of this asterism.

The Party Balloon was the first asterism to start my list. It has a close double star (HU1332) and a multi-double star (BU278) within the Party Balloon's streamer. It is bright, easy to imagine, provides interesting surroundings and is binocular friendly. From this point I searched for more interesting stellar shapes. I was open-minded and imaginative. I knew what I was looking for but unsure of just how many of these pattern asterisms I would find. Not many months into my search I realize that patterns abound. I became choosey, particular and the parameters developed. I soon developed a power point program for school tours at Discovery Park in Safford, Arizona. From that program I was able to utilize the endless imaginations of third and fourth grade school children. And what a large imagination our young people have!

Bowl of Stars

23 hr. 55 min.   +56 deg. 30 min.

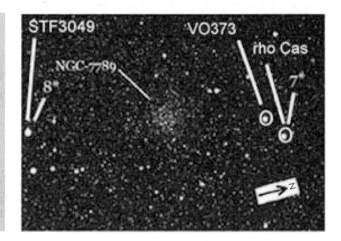

Figure 2.10a.
•Note: Flamsteed
numbering system.
°Note: Variable
stars, Double star.

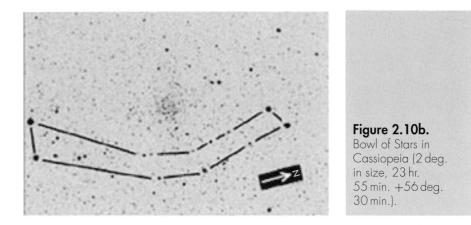

Figure 2.10b.
Bowl of Stars in
Cassiopeia (2 deg.
in size, 23 hr.
55 min. +56 deg.
30 min.).

This open cluster, asterism mix, Fig. 2.10, is among the largest group of patterns in this catalog. At 2 degrees, the bowl that holds this cluster of stars has to be large. NGC 7789 is a faint, compact cluster of stars that requires 3 inches (76 mm) of lens to appreciate. Two bright stars on either end of the bowl designate its boundaries. The cluster is located right in the center of this asterism, floating just above the bowl. The photo shows it all. It is best to observe this mix early in the morning when a bowl full of stars is a great sight for breakfast time. That was a joke. However, you can snack on this object any time of the night with 50 millimeter binoculars at 12 powers. The cluster will be seen as a small faint cloud.

To find this mix, go to star 11 Cassiopeia (Caph) and then 3 degrees westerly. The bowl itself is obvious as morning on the Sky Atlas 2000 with NGC 7789 just above it. 7 Cassiopeia and 8 Cassiopeia depict the bright stars on each side of the bowl. Star STF3049 is a close double with a 7.2 magnitude companion separated by 3 seconds of an arc. The two bright stars at one end of the bowl are variable stars. VO373 displays one-half magnitude changes every 13 days while rho CAS drops 2.1 magnitudes in just under a year. This magnitude change in rho (Flamsteed 7) will take some of the shine off the edge of the bowl. Talking about shine, if your preference is a shiny concave utensil like a reflector, then by all means, check out all of the open cluster's stars in the Bowl of Stars.

Pisces Map and Asterism

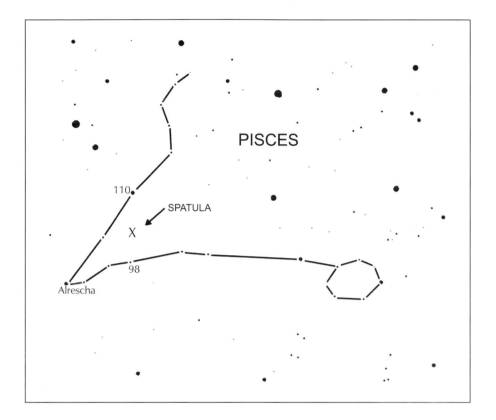

Spatula

01 hr. 35 min.   +08 deg. 00 min.

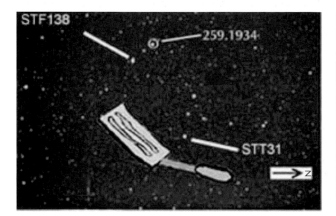

Figure 2.11a.
°Notes: Variable star, Double star.

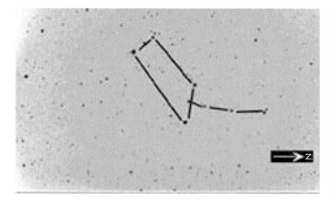

Figure 2.11b.
Spatula in Pisces ($1\frac{1}{2}$ deg. in size, 1 hr. 35 min. +8 deg. 00 min.).

Figure 2.11c.
Spatula in Pisces.

What else would you expect in Pisces the Fishes? Of course, a kitchen implement capable of flipping those fish over in a pan. At $1\frac{1}{2}$ degrees in size, this asterism sure can develop some wrist action. The field is empty of any obstructing background stars and that would be equivalent to a good trout pond void of any stocked fish three days after opening season. The stars in this asterism are bright and form the Spatula shape, Fig. 2.11, most convincingly. Mostly visible in a 50 millimeter finder scope at 10 powers, a 60 millimeter binocular at 16 powers on a mount will show the Spatula asterism completely. On the Mega Star 5 planetarium program, the Spatula shows up at 10.5 magnitudes clearly.

The entire outline is made up of 10 magnitude stars and brighter. The four stars forming the flexible blade sautés in at 6–7 magnitudes. There is a variable star (259.1934) noted on the photo. This variable star will not affect the shape of the Spatula. There are also two double stars on one side of the flexible blade. STT31 is a close split at 4 seconds of an arc and STF138 is a multi-double star. This group of stars would be a great party asterism during a barbecue cook out. The Spatula is very easy to locate between star 110 and star 96 Pisces. The shape of the blade's four stars stands out visibly on the Sky Atlas 2000. On a scale from heads to fins, this asterism is a chef's toolkit delight.

Andromeda Map and Asterism

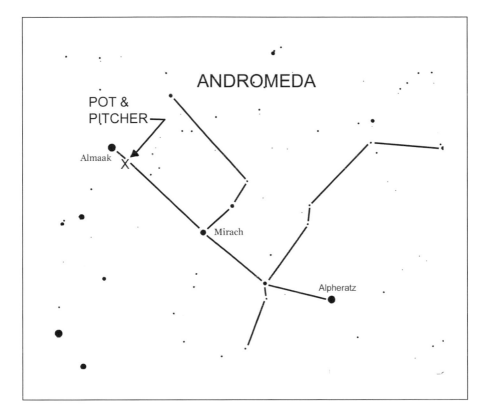

Pot and Pitcher

01 hr. 55 min.   +41 deg. 20 min.

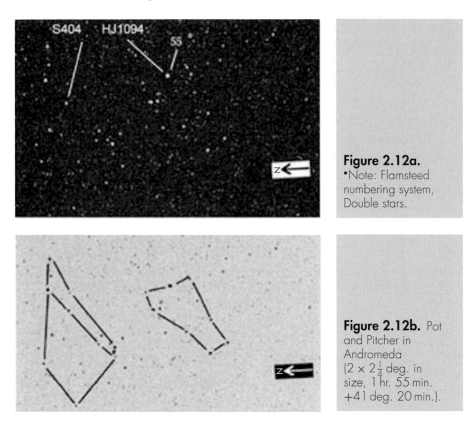

Figure 2.12a.
•Note: Flamsteed numbering system, Double stars.

Figure 2.12b. Pot and Pitcher in Andromeda (2 × 2$\frac{1}{4}$ deg. in size, 1 hr. 55 min. +41 deg. 20 min.).

Here are two asterisms for the price of one in a binocular field of view. A pitcher filled with dark matter and a pot to pour this dark matter into. There are binary stars and now there is a binary pattern asterism, Fig. 2.12. At 2$\frac{1}{4}$ degrees total between the two shapes, we are talking binoculars here. Not to worry, the general image of both asterisms forms nicely around 8.5 magnitudes. Fifty millimeter binoculars at 12 powers will do nicely in a dark sky situation. The pitcher is easy to see, but the pot looks more like a waste paper basket or a foot soldier's hat from muzzle loading days way back when, and requires some imagination. Both Pot and Pitcher are visible on the Sky Atlas 2000 and they are just southwest of Almaak, which is also star 57 Andromeda.

The bright star on top of the pitcher is 55 Andromeda. These two asterisms would be most inviting with coke bottles (large binoculars) since our next door galactic neighbor, the Andromeda Galaxy, is most impressive in binoculars. There is a double star in the Pot and a double star in the Pitcher asterism for reference. S404 is a multi-double star located in the Pot. HJ1094 is located in the Pitcher and has a 5.5 and an 11.0 magnitude stars split by a minute. The Pot and Pitcher are an interesting array of stars that can be rearranged to fit one's imagination. Instead of visualizing a pot pattern, perhaps seeing a pan? This, in my opinion, is what true stargazing is all about.

Ursa Minor Map and Asterism

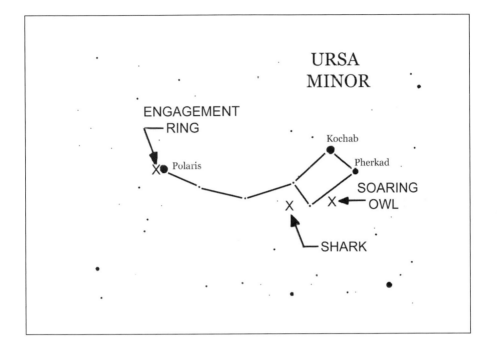

URSA
MINOR

ENGAGEMENT
RING

Kochab

Pherkad

Polaris

SOARING
OWL

SHARK

Engagement Ring

02 hr. 32 min.   +89 deg. 16 min.

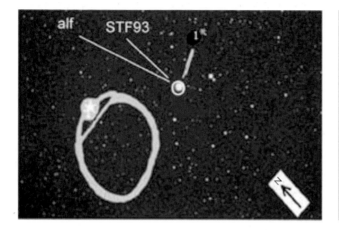

Figure 2.13a.
•Note: Flamsteed numbering system.
°Note: Variable star.

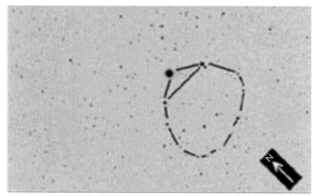

Figure 2.13b.
Engagement Ring in Ursa Minor ($1\frac{3}{4}$ deg. in size, 2 hr. 32 min. +89 deg. 16 min.).

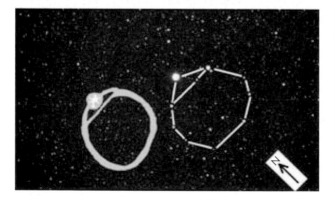

Figure 2.13c.
Engagement Ring in Ursa Minor.

This well-known asterism is as visible as a new diamond ring on the finger of a smiling young lady. Such is the case as this engagement ring, Fig. 2.13, is held out for all to see at the end of the long arm (handle) of the Small Dipper. Like an attention getter, Polaris the diamond vastly outshines the setting this star rests in. Ten faint stars form the ring outline at 10 magnitudes, with the basic shape appearing around 9.5 magnitudes. The two stars on either side of Polaris, somewhat brighter then the rest of the setting, should be considered the baguettes.

At about $1\frac{3}{4}$ degrees, this asterism can occupy a view in many wide field refracting telescopes at low power. Sixty millimeter binoculars with 14–20 powers on a steady mount will no doubt get you into the wedding party. Polaris illuminates at about 2,400 times that of our Sun, making this finger ornament quite a sparkler. The North Star does alter its brightness every four days by 27 tenths of a magnitude. But, alas, there are at least two known suitors (double stars) clutched in Polaris's gravity. Whoa, the sorrows of a love triangle.

Soaring Owl

16 hr. 00 min.   +74 deg. 00 min.

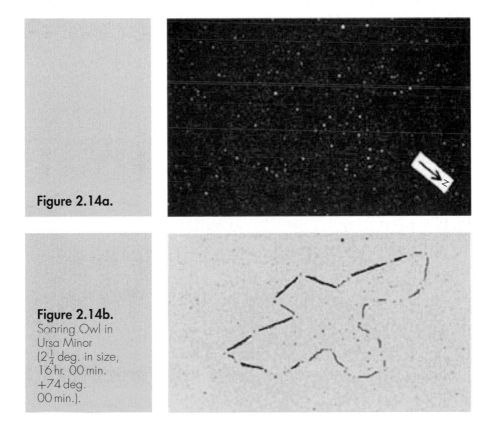

Figure 2.14a.

Figure 2.14b.
Soaring Owl in Ursa Minor $(2\frac{1}{4}$ deg. in size, 16 hr. 00 min. +74 deg. 00 min.).

Most owls don't soar, they swoop. No hot updrafts at night. But this is a catalog of imaginative patterns, so if I say this particular Owl soars, just give me some slack. And, like an owl in the night, Fig. 2.14, this one requires some good night vision to see. This is definitely a large, battleship type binocular object. At least 100 millimeters and up, but don't go over 25 power. Unfortunately, with a $2\frac{1}{4}$ degree wingspan in size, the widest field refractors at the lowest powers are the best choices. Tracing this asterism out in the eyepiece will either raise or ruffle your feathers. This was one of those iffy keep or toss ones. But what the heck, if you can pattern this predator out in the sky, then no small beasties rustling in the weeds are safe at night.

As you can see in the photo, the majority of stars are a faint 11.5 magnitude. The background stars, most of them, only show up on photographs and not in the eyepiece. Once again, the faint stars outline the shape of the Owl. There are two 9 magnitude stars in back of the Owl. One star locates the back corner tail end, but the other star is on the top center of the photo and does not belong to this asterism. These two stars may act as a guide to further locate the Owl. The stars that shape the Owl do stand out in the field of view. The Owl resides only $2\frac{1}{2}$ degrees south of star 17 Ursa Minor. It is difficult to see an owl at night, therefore, patience and a love of flying things will get you through this one.

Shark

16 hr. 45 min.   +77 deg. 50 min.

No lie, this one takes the bite! I have displayed this asterism, Fig. 2.15, on many occasions and it has chased everyone out of the water. You may have seen how sharks circle when it is feeding time. Well, this predator circles around Polaris in the sky. Few people have had trouble seeing this very visible outline in the eyepiece. Quite easy to locate in a pair of 50 millimeter binoculars at 12 powers, this asterism is logged in as a good specimen and measures on the fish scale at the sky dock as $1\frac{1}{2}$ degrees in size.

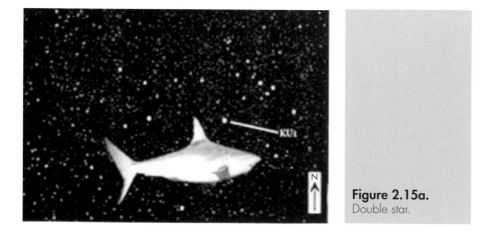

Figure 2.15a.
Double star.

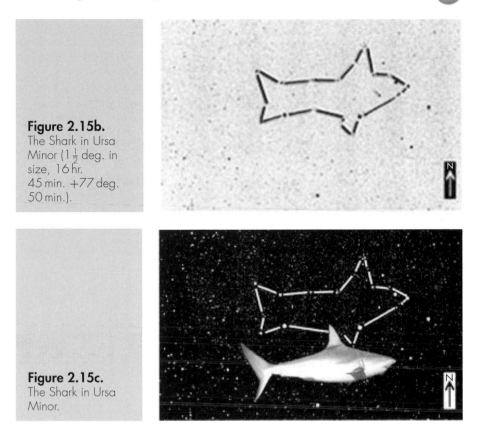

Figure 2.15b.
The Shark in Ursa Minor (1½ deg. in size, 16 hr. 45 min. +77 deg. 50 min.).

Figure 2.15c.
The Shark in Ursa Minor.

Thirteen bright stars stand out well in the field of view. Nine magnitudes show most of the shape and one-half magnitude more reveals that scary fin sticking out of the water.

Twelve stars shape the Shark with the front section looking mighty hungry. The shark is only 2 degrees in a northeast direction from star 21 Ursa Minor. Eleven stars depict the Shark clearly on the Sky Atlas 2000. For those people that go after the faint fuzzes, NGC 1217 floats just above and no doubt keeping very far away from this shark. Planetarium users will find the Shark waiting for them at the coordinates listed above. KU1 is the only multi-star system in this pattern. Since the Shark is located in Ursa Minor, my assumption would be that the Shark has a better chance of successfully tangling with the small cub rather then messing with papa bear.

Perseus Map and Asterisms

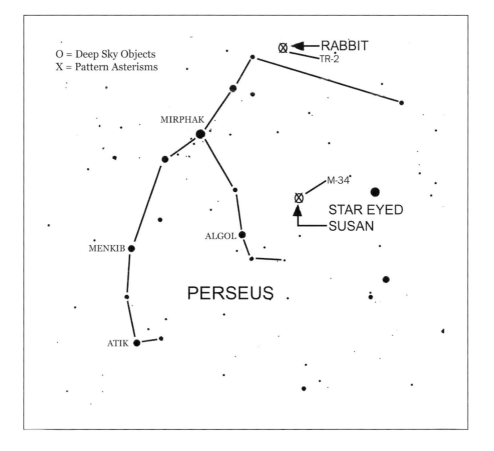

O = Deep Sky Objects
X = Pattern Asterisms

RABBIT
TR-2

MIRPHAK

M-34

STAR EYED
SUSAN

ALGOL

MENKIB

PERSEUS

ATIK

Rabbit

02 hr. 37 min.   +55 deg. 45 min.

Figure 2.16a.
Double star.

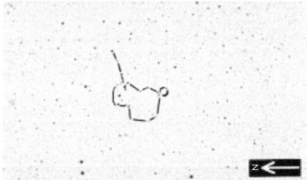

Figure 2.16b. The Rabbit in Perseus ($\frac{1}{2}$ deg. in size, 2 hr. 37 min. +55 deg. 45 min.).

Figure 2.16c. The Rabbit in Perseus.

Trumpler 2 is an open cluster buried in the Milky Way. Photographs of this cluster will not reveal anything resembling a rabbit shape. In fact, the Rabbit asterism, Fig. 2.16, in the TR-2 open cluster washes out very easily in a photo and has to be lightly time exposed to show the rabbit appearance. However, in a telescope, and the bigger the better, the hare pops out of the stellar weeds as if frightened. You will need at least 80 millimeter binoculars and 20 powers to see this little creature. The Rabbit is well camouflaged in the Milky Way, but it is there. Try a carrot, maybe, but a reflecting telescope definitely. Like most (star trek) Vulcan's, the ears give this little fellow away. They stick straight up as this rabbit listens for the slightest movement.

This is another faint pattern open cluster, but the shape shows well against the background stars. The entire cluster is not involved in shaping out this pattern, but the group of stars forming the Rabbit stands out by itself. Eight stars construct most of the pattern at 9.5 magnitudes. Four more stars complete the shape at 10 magnitudes. Some people see this bunny and some will not. With the aid of the catalog photo and clip art, the Rabbit will become visible to more people. Trumpler 2 itself is only 2 degrees west of star 15 Perseus. Pattern open clusters are generally small and the Rabbit being only one-half degree in size is no exception. There is a double star in this pattern which will help planetarium users locate this asterism. I guarantee that once you have found this little stellar outline in your telescope, you will star hop to its location whenever you are in the area.

Star Eyed Susan

02 hr. 42 min.   +42 deg. 30 min.

The Star Eyed Susan, Fig. 2.17, is considered an open cluster, asterism mix. Most open clusters are viewed with 35 powers and higher. That much power reveals only the center of the starry eye of the flower, or M 34. This three-quarter degree shape must be viewed at less than 20 powers with at least 80 millimeter binoculars. Telescopes are good, low F ratios are also good and the less the power the better. Lots of Milky Way

Figure 2.17a.
Double stars.

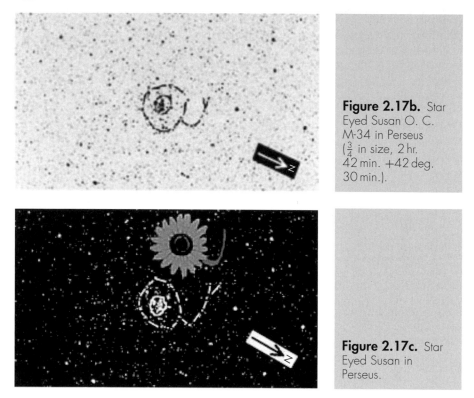

Figure 2.17b. Star Eyed Susan O. C. M-34 in Perseus ($\frac{3}{4}$ in size, 2 hr. 42 min. +42 deg. 30 min.).

Figure 2.17c. Star Eyed Susan in Perseus.

stars in the photo to be sure, but the roundness of the flower and the stem is evident. There are many double stars in this shape. I have pointed out three for location.

Okay, I see a spring flower but you may see a front view of a sliced cucumber, maybe out of a pickle jar? Then again, old memories of a hub cap from your first set of wheels? In any case, the center of M 34, the star eye of the flower Susan, contains stars from 8 to 10.5 magnitudes. The outlining shape of the flower contains nine stars from 8.5 to 10.5 magnitudes. The stem has eight stars and shares the same magnitude as the rest of the flower. The Starry Eyed Susan is only two degrees south of star 14 Perseus and can be ghostly seen in a 50 millimeter finder scope at 10 power. So take another look at M 34 with low power and you may see the same old set of stars in a completely different light.

Eridanus Map and Asterisms

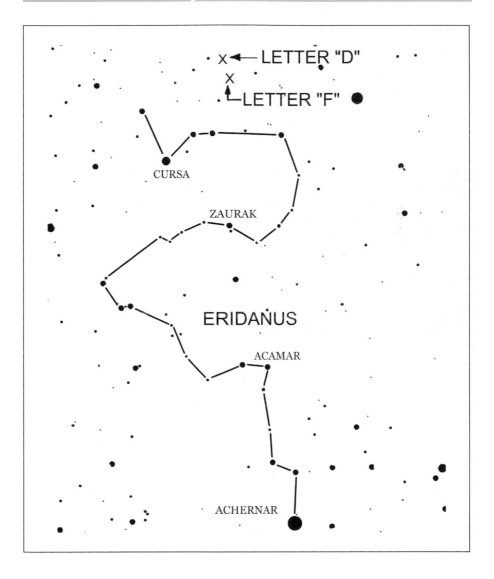

Letter "F"

03 hr. 38 min.   −07 deg. 45 min.

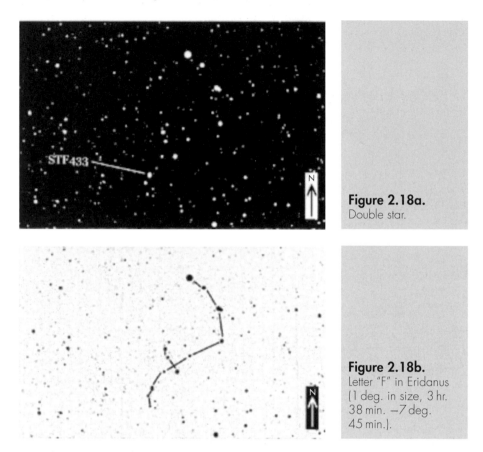

Figure 2.18a.
Double star.

Figure 2.18b.
Letter "F" in Eridanus (1 deg. in size, 3 hr. 38 min. −7 deg. 45 min.).

This letter is just visible in a 50 millimeter finder scope at 10 powers and because of the optical inversion; it looks just like an F, Fig. 2.18, in telescopes. It is one degree in size and if you cannot recognize this asterism, you will need to brush up on your English alphabet. As you can tell from the photo, this is how it would look in a pair of 50 millimeter binoculars at 16 powers. The basic shape starts at 9 magnitudes and 1.5 magnitude more gives you a gold star on your grammar paper. But don't show your star to the student next to you who just saw an F in their telescope. There are a few double stars in this asterism and I plotted one for reference. STF433 has an 8.2 magnitude and an 11.2 magnitude star system separated by over one-half a minute of an arc.

Now, this shape could also look like a sickle with a hand guard or it could be the way some people write the number seven with that little line. You could imagine a grappling hook pattern or maybe big baby tears because someone bent my sword. Okay, I think you get the picture. There are 13 stars in this asterism from the bright

6 magnitude top of the F to the three, 10.5 lower case faint magnitude stars. The F is located just 2 degrees south of star 21 Eridanus. The bright star on top of the letter should have gotten a Flamsteed number, but it failed. There are six stars that show the shape of this asterism quite well on the Sky Atlas 2000. Twelve stars appear at 10.5 magnitudes on the Mega Star 5 planetarium program to bring out this shape perfectly. This particular asterism really does it for those Mmmm, Mmmm, good, Campbell letter soup sippers.

Letter "D"

03 hr. 39 min.   −01 deg. 45 min.

You don't happen to see a pattern here, do you? Well, I was as surprised to find another letter in Eridanus and so close to each other. Anybody happen to have the initials of F.D. or D.F. because this is your constellation. The letter D, Fig. 2.19, is only one-half degree in size and has eight stars from 9 to 9.5 magnitudes. This D is as plain as 'D'ay on

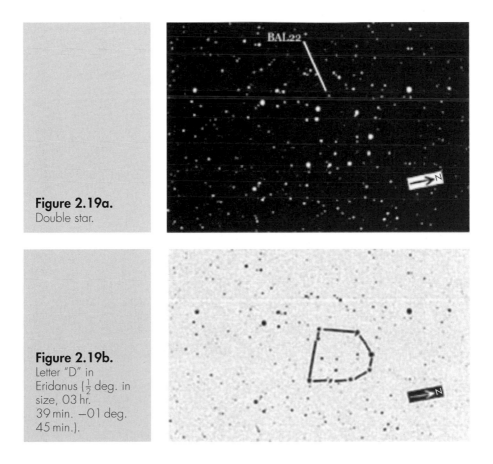

Figure 2.19a.
Double star.

Figure 2.19b.
Letter "D" in Eridanus (½ deg. in size, 03 hr. 39 min. −01 deg. 45 min.).

the Mega Star 5 program at 9.5 magnitudes. The D is one simple-letter easy asterism to find in the sky with 50 millimeter binoculars at 16 powers. Just remember, the D will be flipped for telescope users unless you have a terrestrial prism. After finding the Letter F asterism, it is just a matter of moving north 6 degrees to find the Letter D asterism. There are five stars that show this asterism incomplete on the Sky Atlas 2000, about one degree southwest of star 24 Eridanus.

The D will show up nicely above the background stars. BAL22 is a double star of 10.5 magnitudes. Its companion is 11.5 magnitudes and has a separation of about 5 seconds of an arc. This double is for location purposes only and does not belong to this asterism. There aren't many patterns you can make out of this D shape and I'm sure a half a doughnut will not dunk it. You can turn it sideways and make a mixing bowl or spin it around and make a bright L.E.D. light for checking out your star atlas in the dark. Naturally, the color of the light should be night vision red.

Taurus Map and Asterisms

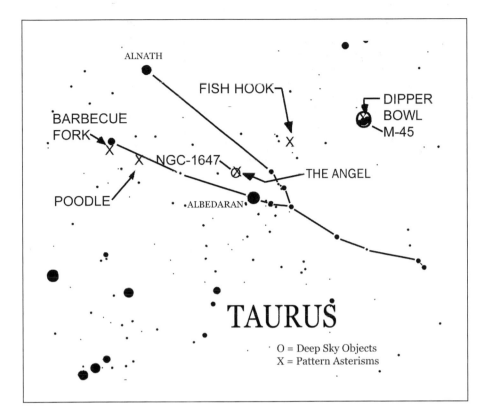

ALNATH

FISH HOOK

DIPPER
BOWL
M-45

BARBECUE
FORK

NGC-1647

THE ANGEL

POODLE

ALBEDARAN

TAURUS

O = Deep Sky Objects
X = Pattern Asterisms

Dipper Bowl

03 hr. 47 min.   +24 deg. 18 min.

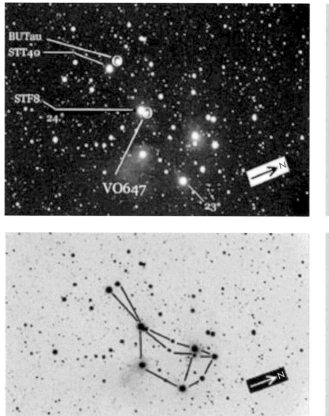

Figure 2.20a.
•Note: Flamsteed numbering system.
°Note: Variable stars, Double stars.

Figure 2.20b.
Dipper Bowl in Taurus (1½ deg. in size, 3 hr. 47 min. +24 deg. 18 min.).

Some non-astronomers see M 45 as The Little Dipper (Ursa Minor) and rightly so because it is small and it looks like a littler dipper. This is, of course, not the case. The Little Dipper asterism is in Ursa Minor and has Polaris as the last star in its handle. It is not difficult to see this little group of stars in almost any skylight condition. That is a good reason why the Pleiades, Seven Sisters, the hair on the bull's back, MEL-22, to mention only a few aliases, makes an excellent pattern open cluster. In binoculars, M 45, Fig. 2.20, really brings home the fact that these handheld little refracting telescopes can do observational astronomy. Most people are impressed considerably when first observing the Pleiades and to find that it is immersed in faint stars when using just common binoculars.

Thirty millimeter binoculars at 8 powers will show the faint stars in this open cluster. On a clear dark night, six or possibly seven stars are visible with the unaided eye in this 1½ degree sized cluster. As expected, this open cluster has been cleanly picked to the solar cores for variable and double stars. The two variable stars I have mentioned dim slightly while the remainder does not affect the dipper pattern. Of the double stars in this cluster, I have also noted just two of them. Because M 45 is a neighboring cluster with bright blue stars, it contains more Flamsteed numbers then a world champion Bingo summit. The Pleiades make an excellent short exposure photo for those beginning astro–camera bugs. In early times, the Pleiades were used to determine seeing conditions at night. Also, one's good eyesight was judged by how many stars they could see on a clear night. So, if you can't find M 45 in the winter season at night and you know where the constellation of Taurus is, it is time for an eye checkup!

Fish Hook

04 hr. 25 min.   +21 deg. 15 min.

This shiny bright asterism will lure a nibble or two from many observers. At $1\frac{1}{2}$ degrees in size it sure can be seen from across the ripples of space. Located just above the face of the bull, I'm wondering if this fisherman's hook hung from a bush hat Taurus adorned when he went fishing, kind of. Fear not, it is my imaginative job to get carried away. There are many bright and interesting stellar shapes in this area to catch one's attention,

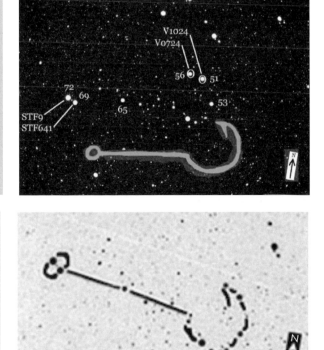

Figure 2.21a.
•Note: Flamsteed numbering system.
°Note: Variable stars, Double stars.

Figure 2.21b. Fish Hook in Taurus ($1\frac{1}{2}$ deg. in size, 4 hr. 25 min. +21 deg. 15 min.).

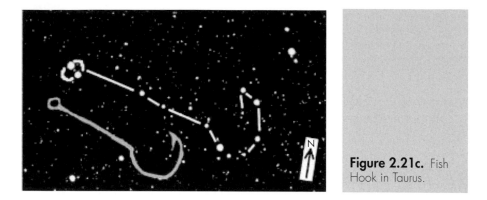

Figure 2.21c. Fish Hook in Taurus.

but I found the Fish Hook, Fig. 2.21, to be simple and very visible. The two bright stars at the end of the hook seem just perfect for tying a fish line to. This area of the sky is just full of interesting star groups for binoculars. Many double stars are located in this area. I picked out just two of them for reference. The two variable stars noted in this asterism do not alter the pattern shape at all.

Sixty millimeter binoculars at 20 powers on a mount will bring out the entire shape of the Fish Hook, but a 50 millimeter finder scope at 10 powers will give a general outline. Seven stars are visible up to 8 magnitudes and at 9.5 magnitudes the Fish Hook is complete. This is a very bright pattern asterism excellent for star parties and very visible on the Sky Atlas 2000. This asterism is just 3 degrees north of the Hyades star group and has more Flamsteed numbers assigned to its stars then a box full of different sized fishhooks. Star 56 Taurus is the sharp end of the hook while stars 72 and 69 Taurus provide the loop for the fish line. Will the Fish Hook be your favorite asterism at the end of your refracting telescope pole? Or will you require just a few more millimeters of wide field eyepiece bate?

The Angel

04 hr. 46 min.   +18 deg. 40 min.

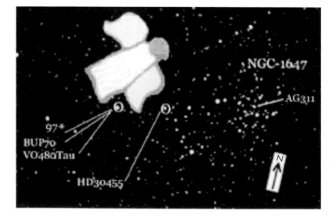

Figure 2.22a.
•Note: Flamsteed numbering system.
°Note: Variable stars, Double stars.

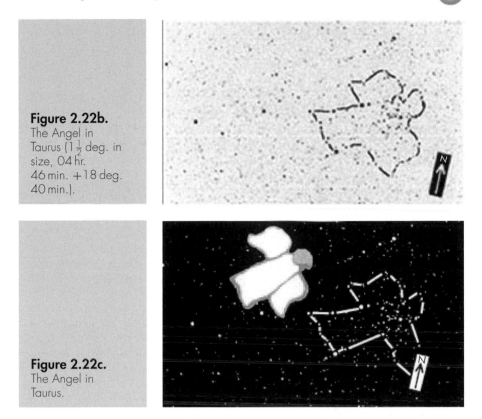

Figure 2.22b.
The Angel in Taurus ($1\frac{1}{2}$ deg. in size, 04 hr. 46 min. +18 deg. 40 min.).

Figure 2.22c.
The Angel in Taurus.

This particular asterism requires a healthy imagination and a leap of faith, to be sure. This is a pattern open cluster mix and NGC 1647 seems to fill the back of The Angel, Fig. 2.22, where the wings are attached with a glow of its own. I have shown this pattern to a bunch of doubting Thomas type people and they could make out only the bottom of the robe. The wings and the head are nothing more then a confusing, ghostly appearance to them. Of course, we are talking about an angel here and angels only appear when you believe in them. As is standard, only the outlining stars are used in tracing out an open cluster pattern. This is where the problem may exist in this shape. Many individuals concentrate on the cluster fill and not on the outer stars. We need to get with the program here people!

The Angel is $1\frac{1}{2}$ degrees in size and the upper part of the body is faint. To find this angel, look $2\frac{1}{2}$ degrees east of the V of the Hyades star group. The star 97 Taurus is only one degree east of The Angel. The Mega Star 5 program will help you see this apparition at 11 magnitudes on your computer monitor. One-hundred millimeter binoculars at 20 powers are needed to envision the faint stars to 11.5 magnitudes. A 5 inch (127 mm) wide field refractor or 6 inch (152 mm) reflector at very low power would just be heavenly. There are certain shapes that would be meaningless without a foundation to build upon. The eight stars outlining the bottom of the robe brought my attention to the rest of the shape, no matter how vague in appearance. How many times have we looked, but yet, have never seen? This Angel will appear in your scope when you least expect it, so be prepared.

Poodle

05 hr. 18 min.   +20 deg. 00 min.

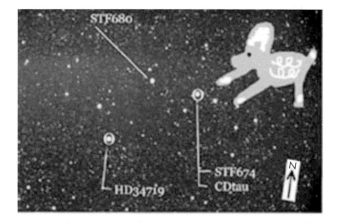

Figure 2.23a.
°Note: Variable
stars, Double stars.

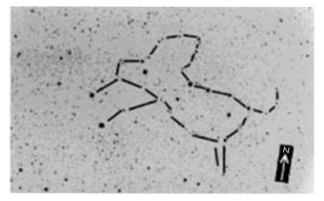

Figure 2.23b.
Poodle in Taurus
($1\frac{1}{2}$ deg. in size,
5 hr. 18 min.
+20 deg. 00 min.).

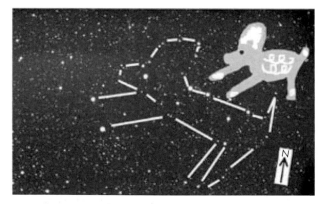

Figure 2.23c.
Poodle in Taurus.

Is the pup in your household the star of your family? I hope so because we have a group of stars that comprise a pup, Fig. 2.23. The Poodles are known for their haircuts and this starlet with a crowning headdress is no exception. This asterism has faint and bright stars and the brighter stars being visible in a 50 millimeter finder scope at 10 powers. Fifty millimeter binoculars at 16 powers will show the faint crown of the Poodle in dark skies. The crown stars are estimated at between 9.5 magnitudes and 10 magnitudes and there are six stars in the crown of the Poodle. This asterism is just $1\frac{1}{2}$ degrees in size and stands out well from the background stars. Two double stars and two variable stars are listed. One star, which just happens to be the neck of the Poodle, contains both groups. The variable star CD tau does have a one-half magnitude change in brightness. This will not affect the pattern.

Is the Poodle easy to see? Star Trek fans should have no problem since it looks like an outline of a poodle materializing in the transporter bay. The back of the Poodle is strung with stars, but the underbelly and bottom of the face is lined with faint stars. A 4 inch (102 mm) refractor at low power brings out the shape well and your imagination must carry on from there. The Mega Star 5 planetarium program collars this pup at 10.5 magnitudes. The darker the sky in your area, the better this asterism will be seen. The Poodle asterism is easy to find in line with the Aldebaran side of the bull's horn, just about in the middle from star 123 Taurus to star 104 Taurus. I rate this asterism as deserving at least one bone-afide viewing at a star party.

Barbecue Fork

05 hr. 43 min.   +21 deg. 20 min.

Just east of star 123 Taurus by about $1\frac{1}{2}$ degrees, this asterism is composed of a string of faint stars. It will require at least 80 millimeter binoculars at 20 powers to view. The Barbecue Fork, Fig. 2.24, is $1\frac{1}{2}$ degrees in length. Being in the Milky Way, there are lots of stars in the field, but the string of stars that shape the Barbecue Fork dots the

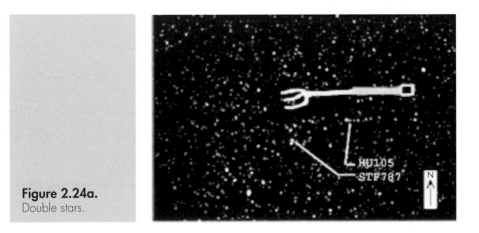

Figure 2.24a.
Double stars.

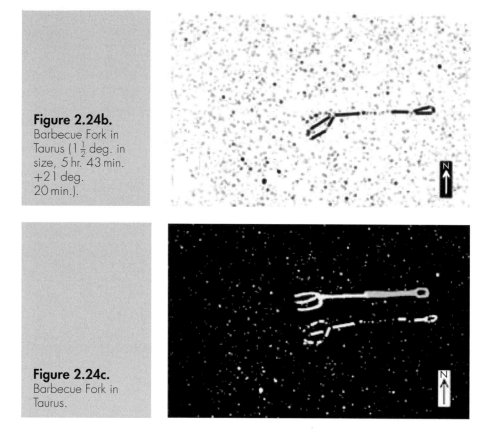

Figure 2.24b.
Barbecue Fork in Taurus ($1\frac{1}{2}$ deg. in size, 5 hr. 43 min. +21 deg. 20 min.).

Figure 2.24c.
Barbecue Fork in Taurus.

area quite well. This was another asterism that could have gone up in stellar smoke, but I decided to test it on a few astronomical chefs first. I made sure I had a clear dark night and seasoned this asterism generously with a 5 inch (127 mm) refractor to help bring out the tantalizing view. No doubt the heat was on and a few onlookers did slice into this asterism with their comments, but overall this outdoor utensil was obvious in the eyepiece. However, not a star is visible in the Sky Atlas 2000.

There are at least 21 stars in this asterism starting at 9 magnitudes and flipping over to around 10.5 magnitudes. The Mega Star 5 planetarium program shows the Barbecue Fork nicely at 11.0 magnitudes to 11.5 magnitudes. There are two double stars in this pattern to give the fork a little flexibility. STF787 is a multi-double star while HU105 is a tight double. No other shape seems to fit this asterism. When I spotted it, a long fork was the first and only image that came to mind. Of course, there is always the chance that I might get barbequed for keeping this asterism in the catalog. I have researched the Barbecue Fork asterism to the point that if I were in England, I'm sure I would hear the chaps say to me, Bravo, well done!

Camelopardalis Map and Asterism

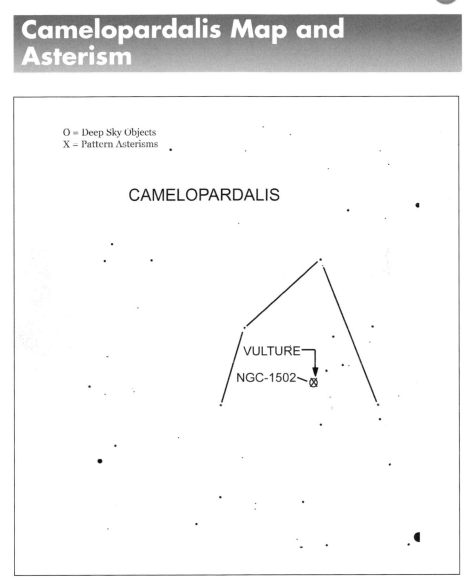

O = Deep Sky Objects
X = Pattern Asterisms

CAMELOPARDALIS

VULTURE

NGC-1502

Vulture

04 hr. 08 min.   +62 deg. 20 min.

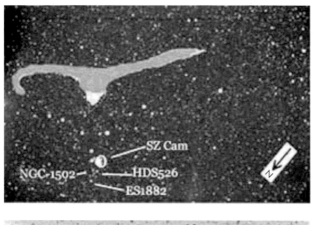

Figure 2.25a.
°Note: Variable star,
Double stars.

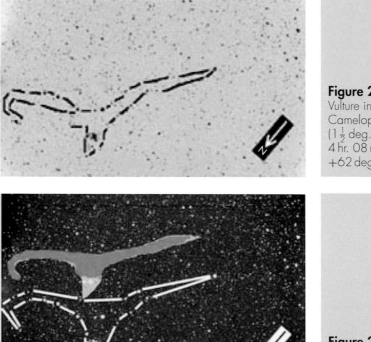

Figure 2.25b. The
Vulture in
Camelopardalis
($1\frac{1}{2}$ deg. in size,
4 hr. 08 min.
+62 deg. 20 min.).

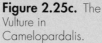

Figure 2.25c. The
Vulture in
Camelopardalis.

There is a string of wavy stars called Kemble's Cascade in Camelopardalis. On one end of the cascade is an open cluster NGC 1502. This cluster forms the beady-eyed face with the pointed beak of the Vulture, Fig. 2.25. The end section of the cascade forms the wings. If you look closely, you would almost think that the right eye of the Vulture is catching sunlight. You might even think this scavenger is winking at you. One wing is curved inward as if this scavenger is circling in for a snack. There are 11 stars that bring out the shape of the Vulture starting at 7.5 magnitudes to 9.0 magnitudes. The Vulture is visible in a 50 millimeter finder scope and the same size in binoculars at 16 powers will show the Vulture nicely. The Sky Atlas 2000 shows the cascading stars with the Vulture's face pointing toward Polaris.

The front star of the beak is 10 magnitudes and a 4 inch (102 mm) refractor at low power will pepper the beak with faint stars. This asterism is just about $1\frac{1}{2}$ degrees in size, depending on just how much wing you wish to give the Vulture. Kemble's Cascade continues for 3 degrees and is obvious when scanning 5 degrees southwest of star 9 Camelopardalis in a 50 millimeter finder scope at 10 power. I have noted two double stars in the beak of the Vulture for reference. The Vulture's eye is indeed a variable star, so it may very well be winking at you with a period of $2\frac{1}{4}$ days. The shape is not difficult to see and the more you look at this bird, the more sinister it becomes.

Orion Map and Asterism

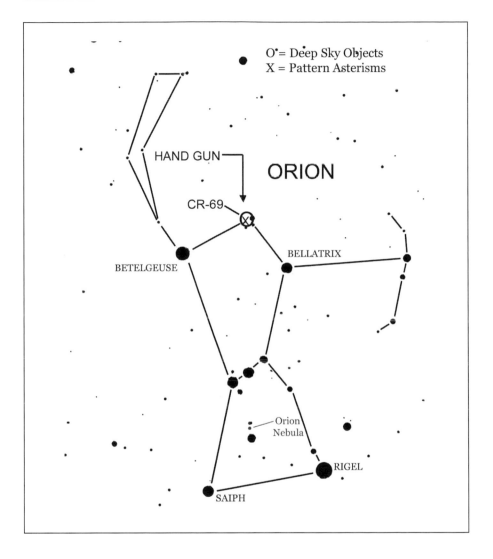

Hand Gun

05 hr. 36 min.   +09 deg. 40 min.

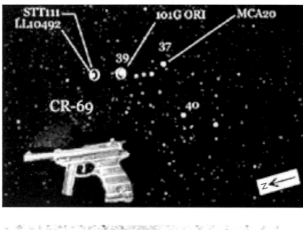

Figure 2.26a.
•Note: Flamsteed numbering system.
°Note: Variable stars, Double stars.

Figure 2.26b.
Hand Gun in Orion (CR-69) (2 deg. in size, 5 hr. 36 min. +9 deg. 40 min.).

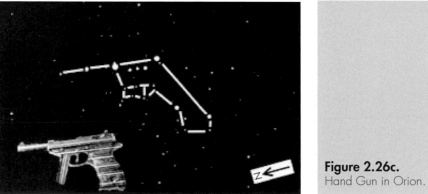

Figure 2.26c.
Hand Gun in Orion.

This pattern open cluster, Fig. 2.26, comes on like gangbusters shooting photons at the speed of light. The very moment Cr-69 moved into my sights I knew I was out gunned. This shape is just so apparent to me, but yet some people cannot see it. In this respect, the clip art does help. A 9 magnitude star starts the north end of the barrel followed by 4, 5 and 6.5 magnitude stars in line. The three stars in a row, about 8 magnitudes, located between the two bright stars, represent the ammunition port. The two next brightest stars are 5 and 7 magnitude and they represent the handle of the gun. Now, if you are accustomed to mowing down lots of bad guys at once, this asterism can also be a machine gun just by extending the handle out to the next star, which would then make it a shoulder stock. The few bright variable stars in CR-69 will not affect the fire power shape of this asterism.

This weapon in Orion's arsenal is 2 degrees in size. After all, we simple can't give this mighty hunter a little peashooter. The remaining parts of this gun involves 9 and 10 magnitude stars, about 10 of them, that form the trigger, trigger guard hand grip and the rest of the handle. CR-69 is very easy to find just above and in between Betelguese and Bellatrix, clearly visible to the eye. In the Sky Atlas 2000 it would certainly be considered a weapon in plain view. Fifty millimeter binoculars at 16 powers will trace (err) out all of the stars in this asterism. If the Hand Gun was pointing almost 180 degrees from its present position, you could say that the Orion Nebula was one of the photon explosions from the Hand Gun. Why not say it anyway since the Orion Nebula is exploding with new stars.

Lepus Map and Asterism

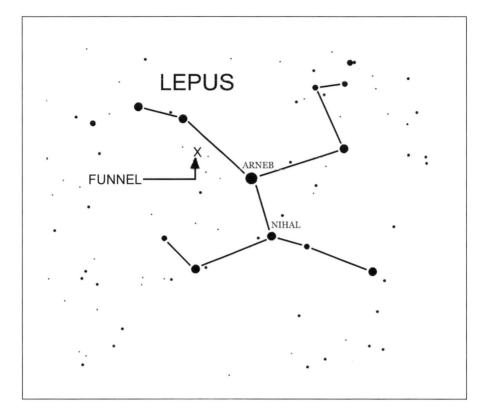

Funnel

05 hr. 46 min.   −15 deg. 40 min.

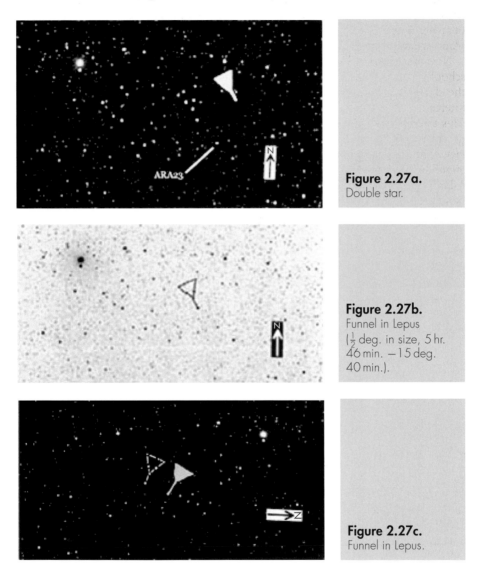

Figure 2.27a.
Double star.

Figure 2.27b.
Funnel in Lepus
($\frac{1}{2}$ deg. in size, 5 hr. 46 min. −15 deg. 40 min.).

Figure 2.27c.
Funnel in Lepus.

Waste not, want not is how the saying goes and the Funnel, Fig. 2.27, will get every drop to wherever you wish it to go. This little one-half degree in size funnel is an asterism worth your time to search for. Six stars pop out at 9.5 magnitudes and the last star in the neck of the Funnel is 10.5 magnitudes. What the Rabbit or Hare is doing with a funnel is unclear. However, if the Funnel had two more stars on either side at the bottom, it would make a perfect Martini glass. Since the Hare is confronted by a hunter (Orion) and his two hunting dogs (Canis Major and Minor) perhaps a stiff Martini in a glass would be in order to calm the little fellow down. Talking about glasses, 2 inch (51 mm) binoculars at 20 powers is acceptable for viewing the Funnel. A double star is noted on the photograph for location, but is not part of this asterism. ARA23 has one 10 magnitude star and one 12 magnitude star separated by 10 seconds of an arc. The shape of the Funnel asterism is displayed perfectly on the Mega Star 5 planetarium program.

Now, if you are a very senior school teacher, the Funnel may look like a hand-held school bell. That kind of bell used in those long ago one room school houses to let the children know school is now in session. The winter constellations boast many interesting objects to view and this asterism will rank among the most tantalizing. This asterism is so visual that including lines or clip art to represent the image isn't necessary, but I did. Because of its size, sizable telescopes can funnel in for a good view. The Funnel is only one degree south of star 14 Lepus. This asterism is faint in a 50 millimeter finder scope at 10 powers. Three stars are visible, just barely, on the Sky Atlas 2000. Oh, by the way, the Funnel can also resemble a plumber's plunger or wisp hand broom. The Funnel is one of my favorite asterisms as well as one of the best for its size and shape.

Monoceros Map and Asterisms

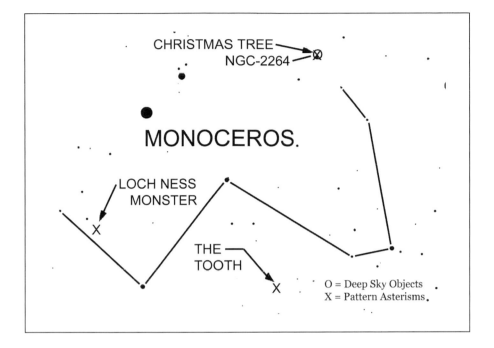

CHRISTMAS TREE
NGC-2264

MONOCEROS.

LOCH NESS
MONSTER
X

THE
TOOTH
X

O = Deep Sky Objects
X = Pattern Asterisms.

Christmas Tree

06 hr. 41 min.   +09 deg. 40 min.

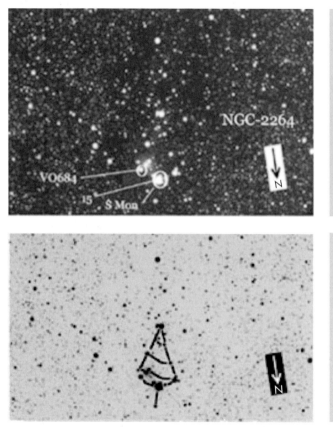

Figure 2.28a.
•Note: Flamsteed
numbering system.
°Note: Variable
stars.

Figure 2.28b.
Christmas Tree in
Monoceros ($\frac{1}{2}$ deg.
in size, 6 hr. 41 min.
+9 deg. 40 min.).

NGC 2264 is a very well-known open cluster with a pattern to boot. Lots of bright stars in this group allow this evergreen to be most events in the stellar forest. This pattern open cluster is a must stop and look for those hobbyists just starting out in astronomy. This cluster of stars is located in the Cone Nebula. The view of this pattern will make up for all of those objects you could not find or see well in your scope. This asterism mix is a nice sight in 50 millimeter binoculars at 12 powers, although more power would be nice. The Christmas Tree, Fig. 2.28, asterism is a wonderful surprise in larger telescopes at low power. The one-half degree tree shape appears at 9.5 magnitudes and if you look hard enough you will see gift wrapped nebulosity under the tree. Finding this open cluster will be distracting because there are so many interesting stellar groups in this area. To locate this stellar pine, find star 24 Gemini, which is Alhena. Go south 4 degrees from Alhena to stars 30 and 31 Gemini and NGC 2264 will be only 3 degrees south of those two stars. It is kind of like going down the chimney right to the Christmas tree, if you're Santa Claus, that is. The brightest star in this pine is Flamsteed number 15. There are lots of double stars and variable stars, to be sure. The two variable stars mentioned on the photo shows little change in brightness and therefore will not alter the tree pattern in any way.

During my presentations on pattern asterisms at Discovery Park, I have received many different ideas from school students. NGC 2264 was viewed as an arrow head. I think that was a third grade group from the San Carlos Apache Reservation. A young lady from another group suggested a tear drop earring. Then there are those very imaginative little scientists in the crowd saying that they see a rocket ship flying among the stars. All of these shapes are acceptable and emphasize the enjoyment and education available with pattern groups of stars in the sky.

The Tooth

06 hr. 54 min.   −10 deg. 20 min.

This asterism will be tough to get a bite on because it is not only faint, but surrounded by the Milky Way. The Tooth is a string of stars from 9.5 magnitudes to 11.5 magnitudes and is one degree in size. There are about 26 stars in this pattern asterism in string groups and evenly spread out in magnitude. As you can see from the photo, the shape does stand out from the stellar crowd. The photograph does show many more stars

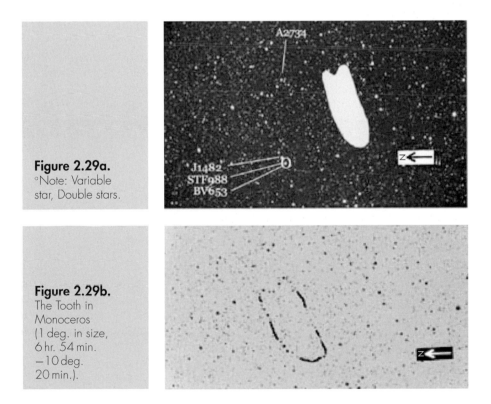

Figure 2.29a.
°Note: Variable star, Double stars.

Figure 2.29b.
The Tooth in Monoceros (1 deg. in size, 6 hr. 54 min. −10 deg. 20 min.).

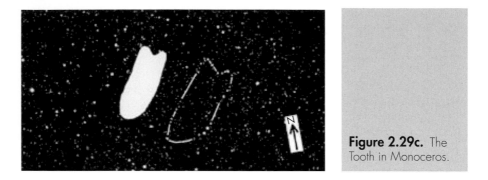

Figure 2.29c. The Tooth in Monoceros.

then will be visible in the eyepiece. The Tooth, Fig. 2.29, is about 2 degrees southwest of M 50, a small but beautiful open cluster. The Tooth requires 80 millimeter binoculars at 16 powers on a mount to view the entire faint shape. A wide field 4 to 5 inch (102–127 mm) refractor or a 6 inch (152 mm) reflector at low power will do very nicely.

I have been told, from a third grader, that the Tooth looks like the bottom of Donald Duck's foot. Another observer suggested that they were looking at a flat–bottomed rowboat from under the water. These images are all acceptable. There are an amazing amount of double stars in the area of the sky. I have pointed out three of them. The single variable star mentioned does not fade noticeably and therefore should not change the asterism. The Tooth is not visible on the Sky Atlas 2000, but you can find it at 10 magnitudes on the Mega Star 5 planetarium program. Finding the Tooth will be a challenge among all of the faint Milky Way stars in the area. However, don't give up on the Tooth asterism because once you have found it, the effort will bring a crowning smile.

Loch Ness Monster

08 hr. 00 min.   −05 deg. 00 min.

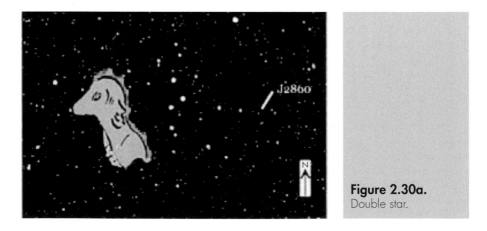

Figure 2.30a. Double star.

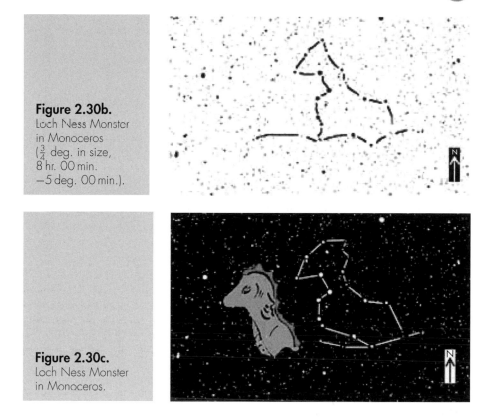

Figure 2.30b.
Loch Ness Monster
in Monoceros
($\frac{3}{4}$ deg. in size,
8 hr. 00 min.
−5 deg. 00 min.).

Figure 2.30c.
Loch Ness Monster
in Monoceros.

Nobody has found this elusive beastie because they have never looked in Monoceros. In that constellation the monster is floating on top of the watery stream of the Milky Way just waiting for anyone to snap a photograph. At first I thought this group of stars represented a rubber duck, but after a while I realized that idea didn't squeak. The Loch Ness Monster, Fig. 2.30, is just visible in 50 millimeter binoculars at 10 powers. The open cluster M 48, the Moth Fig. 2.32 in this catalog, is a little over three degrees in an easterly direction. There are three bright stars in a triangle shape just above the Monster. This triangle group is visible on the Sky Atlas 2000. They are Flamsteed number 27, 28 and 29. The three stars are visible by eye and easy in any finder scope. I see these three stars as a fisherman's net thrown high in the air above the Monster. Hold on world, we will soon have proof!

Eleven stars appear at 9 magnitudes that form the monster's shape. Two more stars pop into this pattern asterism after one-half magnitudes more. Comparing the negative and positive photos, the shape is obvious to image. On the Sky Atlas 2000, the Loch Ness Monster is a small group of eight star just one degree South of star 27 Monoceros. There is a double star that appears to be a drop of water off the monster's back. It contains an 11 magnitude and an 11.5 magnitude pair of stars separated by a little over 5 seconds of an arc. This asterism is three-quarters of a degree in size and is friendly in most medium size telescopes. There are many background stars on the photo that will not appear in the eyepiece. I have imagined hearing one Loch Ness county native tell me on a clear night with his telescope pointing to the sky "The monster is friendly, yes, but don't get too close!" (in magnification).

Puppis Map and Asterism

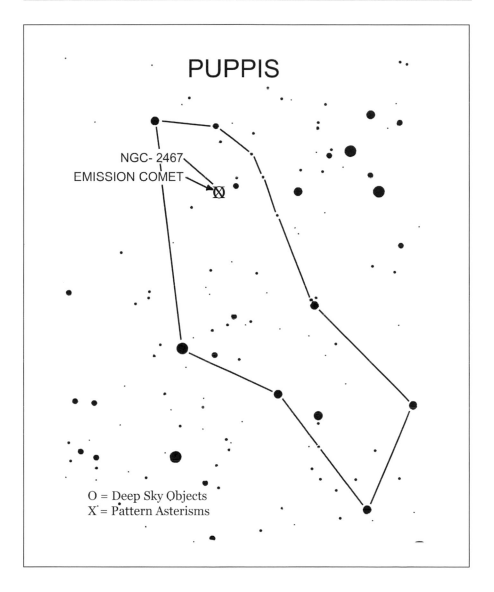

PUPPIS

NGC- 2467
EMISSION COMET

O = Deep Sky Objects
X = Pattern Asterisms

Emission Comet

07 hr. 53 min.   −26 deg. 22 min.

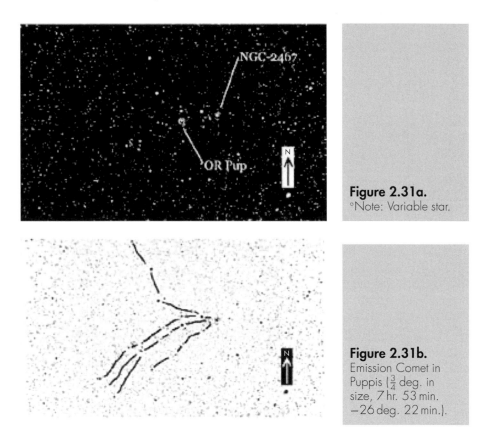

Figure 2.31a.
°Note: Variable star.

Figure 2.31b.
Emission Comet in Puppis ($\frac{3}{4}$ deg. in size, 7 hr. 53 min. −26 deg. 22 min.).

It's a comet! No, it's not a comet and another object goes down on the Messier list. So it was back then when automated telescopes were a thing of the future. Comet hunters are a patient lot, to be sure, but this comet look-a-like can be the jest of the star party. NGC 2467 has that small hazy appearance and the open stars cluster embedded in this emission nebula represents the comet's nucleus. Trailing stars from 10 to 11 magnitudes fan out east of the Emission Comet for three-quarters of a degree, giving that visitor from the Oort Cloud excitement. Best seen in 60 millimeter binoculars at 16 powers for that special effect, medium size telescope will resolve the tail into stars too readily. The Emission Comet, Fig. 2.31, is a far southern object, so check your horizon clearance and humidity layer.

This want to be comet is located almost 2 degrees south of star 7 Puppis. The Sky Atlas 2000 shows eight stars flaring out to the east giving the comet's tail representation. The Emission Comet also gets a little help from open clusters Haffner 18 and 19. They are, by themselves, a minute size hazy concentration of about a dozen stars just north of NGC 2467. The effect of these two smudges makes a little flare up off of the comet's halo. Adding all of these little extras together can sure fool anyone temporarily. So try out this phony baloney comet on an unsuspecting Messier look-a-like using binoculars on one of these winter nights.

Hydra Map and Asterisms

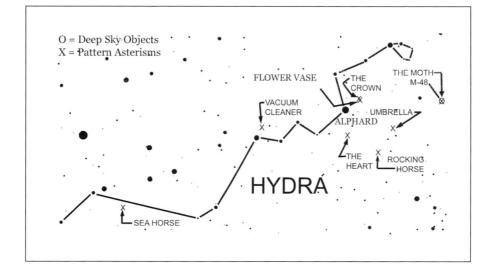

O = Deep Sky Objects
X = Pattern Asterisms

FLOWER VASE

THE MOTH
M-48

THE CROWN

VACUUM CLEANER

UMBRELLA
ALPHARD

THE HEART

ROCKING HORSE

HYDRA

SEA HORSE

The Moth

08 hr. 13 min.   −05 deg. 50 min.

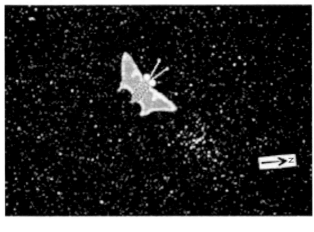

Figure 2.32a.

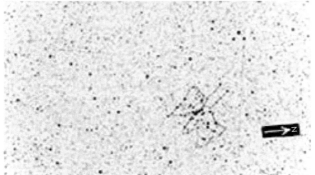

Figure 2.32b. The Moth in Hydra (1 deg. in size, 8 hr. 13 min. −5 deg. 50 min.).

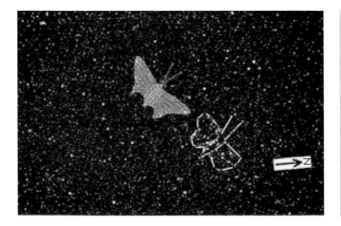

Figure 2.32c. The Moth in Hydra.

The truth is, M 6 the Butterfly cluster, is already taken, so we check my wallet and pull our M 48, the Moth, Fig. 2.32. M 48, the open cluster, is quite a showpiece in the sky. Now it deserves the distinction of being a pattern open cluster called the Moth. Checking my wallet I noticed that the Moth also enjoys these new monetary bills the U.S. mint has produced. The pattern of the Moth flies out at you around 12 magnitudes and covers about one degree of sky. Eighty millimeter binoculars at 14 powers will show the Moth. Because of its size, refracting telescopes and low focal length, medium size reflectors are invited to chase the Moth around the sky. Please note that a wide field of view is recommended to see the shape of the Moth. The Moth is visible in any finder scope under dark skies as a chunk of dust in the Milky Way's starry closet; no doubt the remains of my summer sweater.

The Mega Star 5 planetarium program shows The Moth very well at 10.5 magnitudes only. To finding M 48, locate stars 1 and 2 Hydra on any good star atlas. The Moth is about 3 degrees southwest from those two stars. Like all open clusters used to make patterns in this catalog, the outer stars line out the shape only. On this particular shape, the clip art helps reveal the Moth's outline. Twenty-five degrees north of the Moth is M 44, the Closed Flower. I do hope the Moth is making its way up there to feed because, quite frankly, it has grown accustomed to my material wealth!

Umbrella

08 hr. 40 min.   −12 deg. 30 min.

I have seen people using umbrellas when the Sun was out to protect them. I have my trusty sunglasses and Marvin the Martian bush hat instead. This particular umbrella is for those really very sensitive to starlight folks because you can't see the Umbrella pattern asterism, Fig. 2.33, when it is raining at night. Like a sunny day, the Umbrella is

Figure 2.33a.
•Note: Flamsteed numbering system, Double stars.

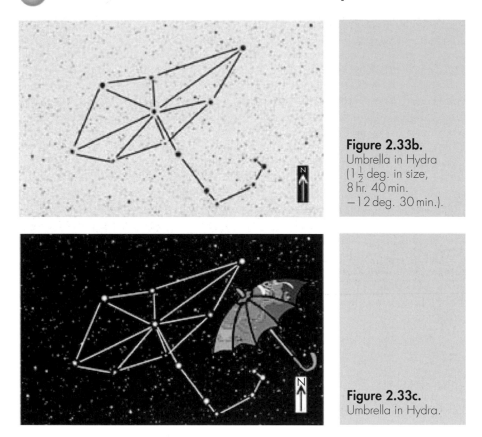

Figure 2.33b.
Umbrella in Hydra
($1\frac{1}{2}$ deg. in size,
8 hr. 40 min.
−12 deg. 30 min.).

Figure 2.33c.
Umbrella in Hydra.

bright and easy in 50 millimeter binoculars at 10 powers. This umbrella is a good $1\frac{1}{2}$ degrees in size, just large enough for two observers and a wide field telescope. The Umbrella is classified as a top notch pattern asterism because of its size, brightness and shape. Why, just the cane on the end of the handle gives this wet weather accessory its identity. Unfortunately, double star number STF1261 represents some wind damage, or perhaps I should say gravitational disturbance, bending the shape of this umbrella.

At eight magnitude, nine stars lineout this asterism well and a half magnitude more of stars stops most of the photons from above. The brightest star in the staff of the Umbrella is star 6 Hydra, an easy find in the sky. On the Sky Atlas 2000, the entire Umbrella is fully open to view. This asterism is only 9 degrees southeast of M 48 and contains 14 stars total. There are three double stars in this asterism for location. RST3603 is a tight system and STF1260 is a close double. However, STF1261, the bent rib in the Umbrella, has a 7.9 magnitude and a 10.6 magnitude pair separated by one-half minute of an arc. This asterism is a most excellent show and tell pattern at star parties and that's no bumbershoot.

Rocking Horse

08 hr. 58 min.   −16 deg. 30 min.

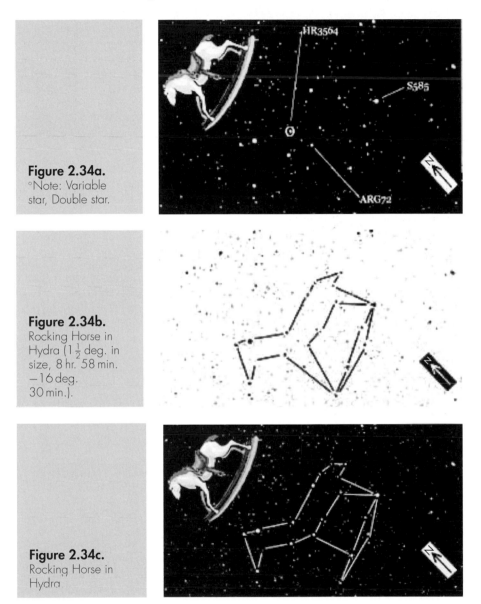

Figure 2.34a.
°Note: Variable
star, Double star.

Figure 2.34b.
Rocking Horse in
Hydra ($1\frac{1}{2}$ deg. in
size, 8 hr. 58 min.
−16 deg.
30 min.).

Figure 2.34c.
Rocking Horse in
Hydra

Who has never had a rocking horse, Fig. 2.34, when they were young or something
that rocked back and forth and was called a rocking horse? There are 13 stars to 8.5
magnitudes that shape up to be just that, a rocking horse. Add a few fainter stars to
comb out the mane and we are ready to ride! I spotted this asterism in the eyepiece and

I yelled "Yeeha!" Actually, it was more like oh wow! But I had to pad my part, in this case for the bottom of the saddle. The Rocking Horse is clearly visible in 50 millimeter binoculars at 10 powers. There are a few background stars but this bright image is easy to pick out using the photo example and the clip art. This asterism is $1\frac{1}{2}$ degrees in size and another excellent bet at any racetrack. Locating this green broke Philly is really easy during feeding time, just 3 degrees east of star 9 Hydra. If you have just finished looking at the Umbrella, all you need do is go 6 degrees southeast and you are ready to rock.

There is a variable star in the Rocking Horse right about where the mane area is and that is a great securing handle for balance when riding. HR3564 does not vary enough to change the shape of this bronco. Two double stars are involved in this asterism. S585 has a 5.8 magnitude star and a 7.0 magnitude star separated by a little over one minute of an arc. ARG72, on the other hand, is a very close double star. You will find about eight stars from the Rocking Horse visible on the Sky Atlas 2000 and all of them will be located in the corral section of Hydra, of course.

The Crown/Flower Vase

09 hr. 04 min.   −04 deg. 05 min.   09 hr. 05 min.   −04 deg. 15 min.

Two for the price of one or a psychiatrist's special. It was only a matter of time I would lose my mind and see two asterisms in the same pattern at different angles. I found the Crown asterism, Fig. 2.35, one year and the second year I found the Flower Vase asterism, Fig. 2.36. When I started compiling this catalog I realized that the coordinates were too close to call. There was only one thought that came into my mind at that point; vacation time! But, after close examination of the patterns and not my burnt bacon brain, it hit me like Tinker Bell dust in front of my eyes. I had photographed the same asterism and had envisioned two different shapes. Now, this is quite different than making up a few shapes out of one asterism. I'm a firm believer that the shape of a pattern asterism is in the eye of the beholder. These asterisms are superimposed and I refuse to separate them or toss one out. Instead, I will examine each as an individual and let the masses mess with it.

The Crown

09 hr. 04 min.   −04 deg. 05 min.

The Crown or Tiara of stars, Fig. 2.35, requires a good taste in exterior cranial fashion. The entire Crown is composed of 8–10 magnitude stars and is visible in 50 millimeter binoculars at 10 powers. The two stars at the very top of the Crown are the only exceptions at 7 magnitudes and 7.5 magnitudes. The entire Crown is approximately $1\frac{1}{4}$ degrees in size and there are plenty of faint stars in the eyepiece. There are about 20 stars in this asterism, depending upon how many jewels are inserted to satisfy your fancy. A 3 inch (76 mm) wide field refractor at 16 powers might help, but it all depends on your healthy imagination. It may require checking out the photo and the clip art a few times, but if you can see this asterism without much assistance, why, it would be a crowning achievement.

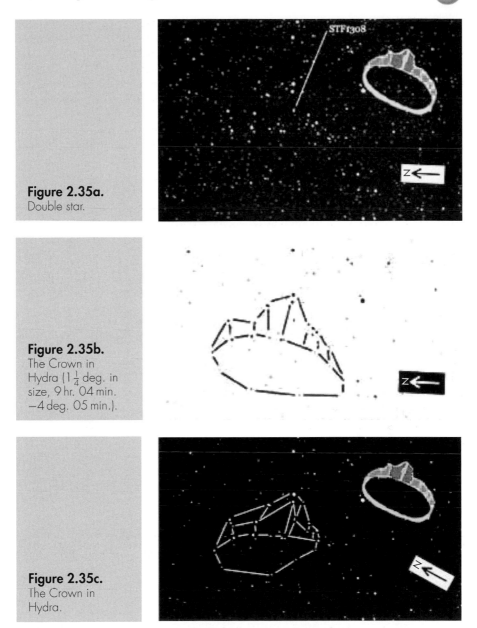

Figure 2.35a.
Double star.

Figure 2.35b.
The Crown in
Hydra (1¼ deg. in
size, 9 hr. 04 min.
−4 deg. 05 min.).

Figure 2.35c.
The Crown in
Hydra.

Star 30 Hydra, Alphard, would be the first star to locate when starting to search for this asterism. Star 23 Hydra, 4 degrees northwest from Alphard is our second point star. That same distance northwest, four degrees that is, should put you on the Crown asterism. Three stars are visible on the Sky Atlas 2000. These three stars are the center rubies of the Crown. The most southern star is the double star STF 1308. This asterism is the only pattern in this entire catalog that will give you a second chance. If the Crown asterism does not do it for you, perhaps the next asterism located in the same space will.

Flower Vase

09 hr. 05 min.   −04 deg. 15 min.

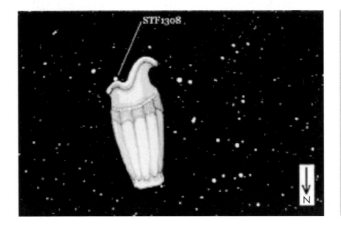

Figure 2.36a.
Double star.

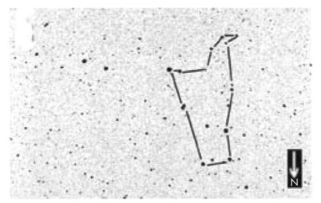

Figure 2.36b.
Flower Vase in
Hydra (1 deg. in
size, 9 hr. 5 min.
−4 deg. 15 min.).

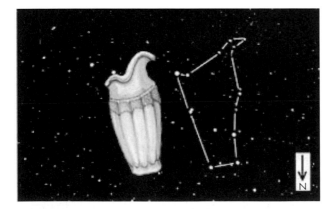

Figure 2.36c.
Flower Vase in
Hydra.

The Flower Vase, Fig. 2.36, is the second asterism mingled within this group of stars. If you take a close look at the Flower Vase, then tilt your head 90 degrees to the left very carefully, you will see the Crown. The clip art is in the way a bit so be forgiving. This photo shows the stars brighter then the first photo, (of the Crown) although the film type and timed exposure was the same. It could have been a very clear night or fresher developing chemical at the photo store? It really hurts when I think about it, so I don't. The Flower Vase is easy to see compared to the Crown. The Flower Vase is one degree in size. It has 10 stars of 8.5 magnitude or brighter that trace out the shape very well. There are a total of 12 stars in this asterism.

The Flower Vase is very visible in the Mega Star 5 planetarium program at 9.5 magnitudes. STF1308, the only visible double star, has an 8 magnitude and a 9.2 magnitude star separated by $10\frac{1}{2}$ seconds of an arc. There is an interesting pair of close together stars on the left side of the vase half way up the container. At first I thought this would be a double star system, but they are just in our line of sight. There is one other point of interest about the Flower Vase. There are no flowers in this Vase, as you can see. The Flower Vase is upside down in the sky and thus all of its contains emptied into the southern hemisphere. This duel asterism is a must for those "I can see anything you can show me" kind of individuals.

The Heart

09 hr. 43 min.   −13 deg. 50 min.

Eight stars show the basic shape of The Heart, Fig. 2.37, at 10.0 magnitudes. Sixteen stars down to 11.5 magnitudes complete this asterism. Get out those battleship binoculars and a sturdy mount. A 4–6 inch (102–152 mm) telescope of your choice at low power would be best. Love is blind and this Heart is faint, but it is there. This pattern asterism was a toss up to stay or go, but I had a good feeling about this one, so I engaged it into this catalog. I must confess that there were certain patterns I really wanted to find in the sky and a heart pattern was one of them. My daughter Altaira

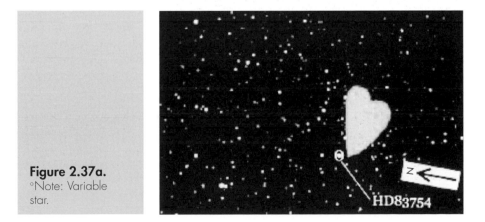

Figure 2.37a.
°Note: Variable star.

HD83754

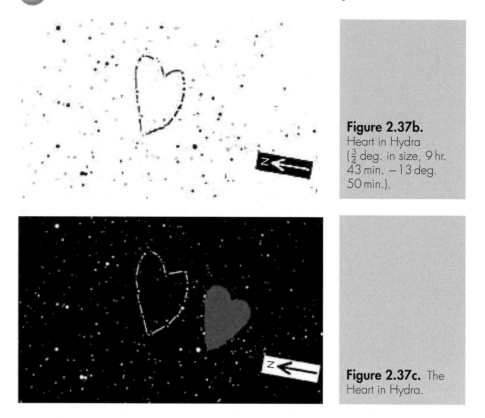

Figure 2.37b.
Heart in Hydra
($\frac{3}{4}$ deg. in size, 9 hr.
43 min. −13 deg.
50 min.).

Figure 2.37c. The
Heart in Hydra.

wanted me to find a Ferret in the sky. A beady eyed long furry hot dog that can walk backwards around corners, go in your shirt and out your sleeve and turn completely around in a winter sock. I found half a Ferret, but the back portion was missing. Not surprising for the antics of a Ferret.

The Heart is just northeast of star 38 Hydra by less then a degree. There are lots of faint stars in the area but most will not show up in the eyepiece. The Heart does a fair job of dotting its pattern in the eyepiece, although there seems to be a few stars that were camera shy. There is a 6 magnitude variable star numbered HD83754 that is a little over one degree north of 38 Hydra. The variable star is not involved in this asterism, but will assist in a lonely heart's club hunt. There are only two stars visible on the Sky Atlas 2000 that belongs to The Heart asterism. The Heart is three-quarters of a degree in size, so it is a small heart that requires a lot of patience and understanding.

Vacuum Cleaner

10 hr. 48 min.   −15 deg. 10 min.

This pattern asterism gets the good housekeeping seal of approval for just the right size, brightness, power, visibly shaped and ease of handling in the Hydra storeroom.

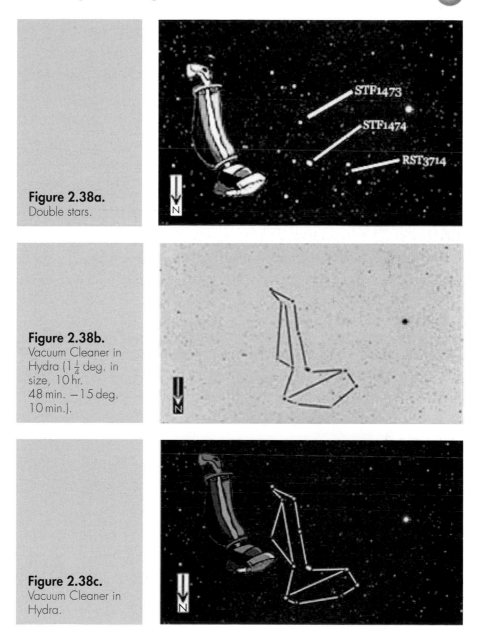

Figure 2.38a.
Double stars.

Figure 2.38b.
Vacuum Cleaner in Hydra (1$\frac{1}{4}$ deg. in size, 10 hr. 48 min. −15 deg. 10 min.).

Figure 2.38c.
Vacuum Cleaner in Hydra.

The angle appearance is like looking up at the Vacuum Cleaner, Fig. 2.38, from below a spotless glass floor. What a commercial that would make! Fifty millimeter binoculars at 10 powers are all that is needed to find this labor saving device hanging upside down in the sky. Plug it in to your telescope's battery pack with the extension cord provided and clean up all the obscuring dust in the Milky Way. There are eight stars at 9 magnitudes or brighter that line out the shape. Three or four more stars down to 10.5 magnitudes complete the Vacuum Cleaner. The Mega Star 5 planetarium program reveals this group of stars at 9.5 magnitudes the best. A really good star party pattern asterism, but give out the details first before the observing begins.

To find the Vacuum Cleaner, just look for star V Hydra. That is right on the boarder of Crater and the Vacuum Cleaner is just north by a degree. Seven stars are clearly visible on the Sky Atlas 2000. I have three double stars listed for reference. STF1473 is a multi-double in this asterism. It is located right about where the on/off switch would be on most vacuums and that is a good place for a double star. STF1474, is an interesting multi-double easy to split in the motor housing of the Vacuum Cleaner. The last double, RST3714, is a tight pair to split at $1\frac{1}{2}$ seconds of an arc. Once you have collected up all of the star party viewers with the Vacuum Cleaner, remember to empty the vacuum bag in the center of our Milky Way Galaxy. Our resident black hole will be more then happy to dispose of the material.

Sea Horse

13 hr. 10 min.   −23 deg. 30 min.

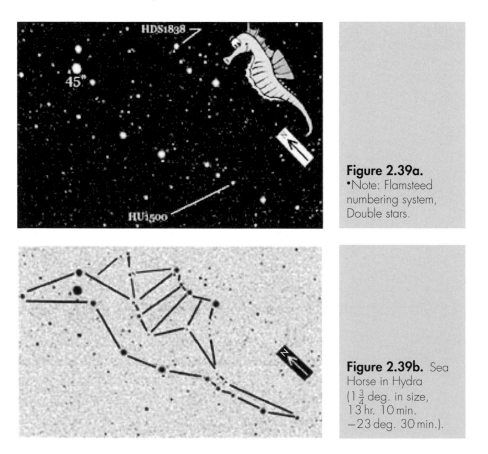

Figure 2.39a.
•Note: Flamsteed numbering system, Double stars.

Figure 2.39b. Sea Horse in Hydra ($1\frac{3}{4}$ deg. in size, 13 hr. 10 min. −23 deg. 30 min.).

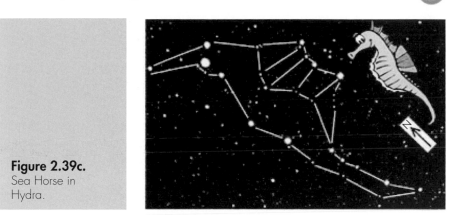

Figure 2.39c.
Sea Horse in
Hydra.

Now this asterism is an interesting species. I think little sea horses are one of nature's wonders. However, the Sea Horse, Fig. 2.39, in Hydra is giving me that look like, who do you think you're kidding, buddy! When I spotted this pattern of stars the first idea that came to mind was a sea horse. But where is the curly tail, I thought? Oh, I almost forgot, this is a thoroughbred racing sea horse and they rarely curl their tails. With that said and I know I didn't get away with it, let's continue. This little sea horse is not little by any means. Mr. Speedy, who races at salty kelp downs, measures in at $1\frac{3}{4}$ degrees in length. A standard 50 millimeter binocular at 10 powers gets a great view from the grandstands. Thirteen stars shape out the Sea Horse at 9 magnitude and much brighter. A few more stars appear at 10 magnitudes to fill in the pattern, but are rarely needed. Whether you envision a sea horse or a stealthy sunfish makes no matter, this is a wonderful group of stars to view through the glass of any aquarium equipment refracting telescope.

The Mega Star 5 planetarium program shows this sea creature at 10.5 magnitudes perfectly. Finding the Sea Horse is as easy as locating star 46 Hydra and the two sea horse eyes will be only 2 degrees west, looking your way. One of the bright horsy eyes is logged in as Flamsteed number 45. The Sea Horse is just on the north border of Hydra and Virgo, no doubt exercising for the next race. There are two close double stars located on Mr. Speedy, one on the upper fin and the other in the tail. There was one suggestion made during a presentation by a fourth grade marine biologist. Children at that tender age know exactly what they are going to be when they grow up. If you ignore the fins on the back of Mr. Speedy, then you would have a nasty looking alligator. Watch out Captain Hook!

Cancer Map and Asterisms

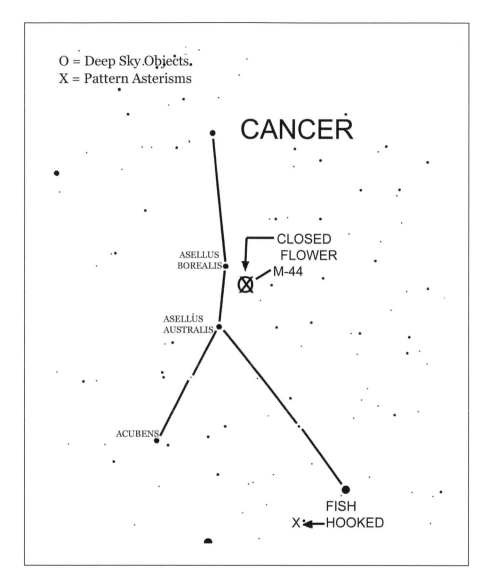

O = Deep Sky.Objects.
X = Pattern Asterisms

CANCER

CLOSED
FLOWER
M-44

ASELLUS
BOREALIS

ASELLUS
AUSTRALIS

ACUBENS

FISH
X◄─HOOKED

Fish Hooked

08 hr. 28 min.   +07 deg. 20 min.

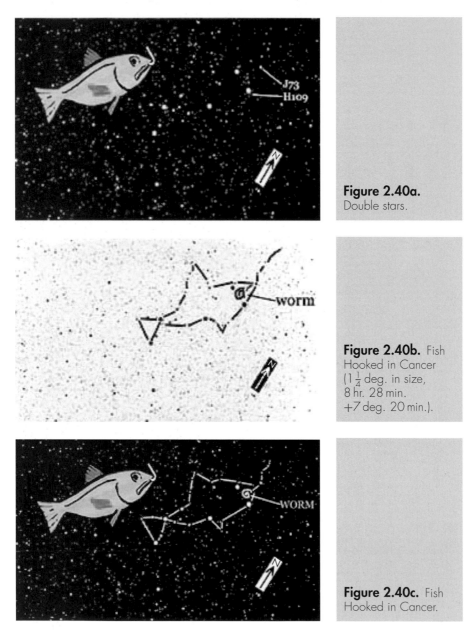

Figure 2.40a. Double stars.

Figure 2.40b. Fish Hooked in Cancer ($1\frac{1}{4}$ deg. in size, 8 hr. 28 min. +7 deg. 20 min.).

Figure 2.40c. Fish Hooked in Cancer.

Now don't get confused between the Fish Hook in Taurus, Fig. 3.3, a device used to catch a fish, and the Fish Hooked in Cancer, Fig. 2.40, which is this device already in use. This asterism is composed of bright and dim stars, but the basic outline is bright enough to be seen in 50 millimeter binoculars at 16 powers on a mount. Eighty millimeter binoculars will show the worm in the mouth of the fish and the fish, in this case, looks like Charlie the Tuna. The group of stars making up the worm is 12 magnitudes. In the photo, Charlie the Tuna looks like he's on his way up to the surface where he will become vacuum-packed. There are a lot of faint stars in this area which do not interfere with the asterism's shape. Twelve stars at 9.5 magnitude and brighter form the fish shape and two stars in that group make up the fish line. The fin on top of the fish appears at 10 magnitudes.

There are eight stars on the Sky Atlas 2000 that represent the image of this fish very nicely. The entire pattern asterism is $1\frac{1}{4}$ degrees in size and is on the border with Hydra. The Fish Hooked is only 3 degrees southeast of star 17 Cancer. Very interesting to see is star 17 Cancer and the three stars next to it. On the Sky Atlas 2000 they almost look like the top view of a fisherman securing his catch, and in this case it's Charlie the Tuna. The double star H109 is the brightest star in this pattern and in an area of the fish that must really hurt the most. The second double star, J73, is the opening of the fish's mouth. This pattern asterism is an easy catch when fishing in the stellar pond, as someone, optically equipped, has already found out.

Closed Flower

08 hrs. 41 min.   +19 deg. 30 min.

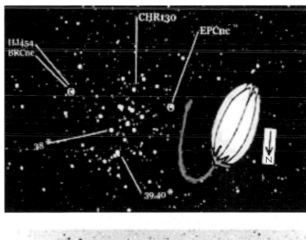

Figure 2.41a.
•Note: Flamsteed numbering system.
°Note: Variable stars, Double stars.

Figure 2.41b.
Closed Flower in Caner ($1\frac{1}{2}$ deg. in size, 8 hr. 41 min. +19 deg. 30 min.).

Figure 2.41c.
Closed Flower in Cancer.

This hazy spot in the sky has been seen by the ancients and called the babe in the manger. They could have envisioned a mattress of hay supported by three surrounding stars. On a clear dark night looking up into the constellation of Cancer the Crab, such a sight is not too difficult to imagine. In more recent times it has taken on the name of the Beehive or Praesepe (Manger) cluster. Some familiar open clusters take on a new appearance in binoculars at 10 powers than in a telescope at 35 powers. Their visual shape suddenly blossoms into stars as well as patterns from a distance. In fact, M 44, Fig. 2.41, looks just like a closed flower, and being called the Beehive is really not good news for those pollen gathers, that's for sure!

This large, irregular open cluster is $1\frac{1}{2}$ degrees in size. It is located between star 43 Cancer and star 47 Cancer, which is South of M 44. There is a gathering of Flamsteed numbers surrounding the manger as if they were those pesky sheep on a Serta mattress commercial. HJ454 is a faint double and CHR130 is a very tight double star system. There are a lot of variable stars in this open cluster but not much data about their fluctuation. Stars to 9.5 magnitudes will show the Closed Flower shape well in 50 millimeter binoculars at 10 powers. I could easily imagine one bee buzzing to the next; "It would really be nice to see M 44 open for business." Unfortunately, the majority of flowers close at night, the only time you can see this open cluster.

Lynx Map and Asterisms

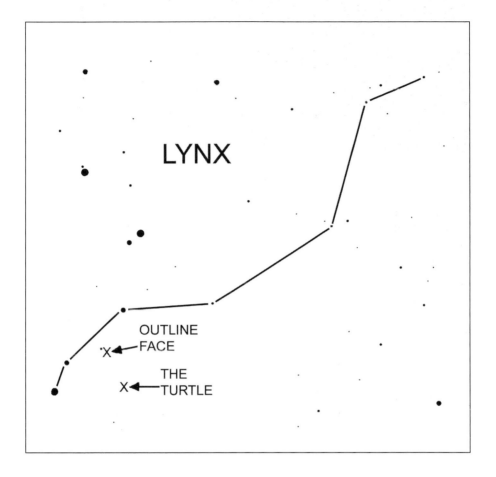

LYNX

OUTLINE
X← FACE

THE
X← TURTLE

The Turtle

08 hr. 54 min.   +35 deg. 00 min.

Figure 2.42a.
Double stars.

Figure 2.42b. The Turtle in Lynx ($1\frac{1}{4}$ deg. in size, 8 hr. 54 min. +35 deg. 00 min.).

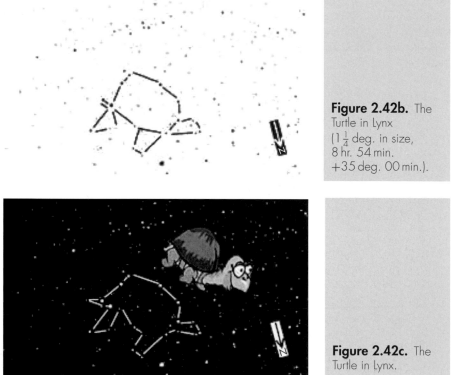

Figure 2.42c. The Turtle in Lynx.

Do you have a camper, trailer or maybe a fifth wheel lugging behind you as a travel companion? Your roof, enclosure and security, just like a turtle. It sure comes in handy when there are wild animals around like the Lynx. The shell of The Turtle, Fig. 2.42, is made up of 9 magnitude stars with the legs and outstretched head to 10.5 magnitudes.

There are about 18 stars in this pattern asterism. The shell is easy in a 50 millimeter finder at 10 power and a 60 millimeter binoculars at 16 powers on a mount will resolve the entire turtle. There are lots of faint stars in the area that do not present a shell game at all. The Turtle is $1\frac{1}{4}$ degrees in size and is located just 3 degrees north of star 59 Cancer, which is just over the border from the Lynx.

I picked out two double stars that are located at the tail and neck of the Turtle. STF1272 has an 8 magnitude and a 10.2 magnitude pair separated by a little over 12 seconds of an arc. COU1894 is a tight double star. Nine stars are visible on the Sky Atlas 2000 as this homebody hangs upside down in the sky. Clip art placed next to this group of stars will help sensitize the imagination visually. The shell is easy to spot in the sky and even looks like a letter C. As you may know, box turtles are very shy, so remain motionless by your eyepiece, briefly, as the head and legs will slowly appear out into view.

Outline Face

09 hr. 05 min.   +38 deg. 20 min.

Here's looking at you kid takes on an entirely new meaning when a face full of stars is staring back at you. The Outline Face, Fig. 2.43, is composed of bright and faint stars. The big nose is visible in 50 millimeter binoculars at 10 powers. Eighty millimeter binoculars at 16 power "at the steady" (mounted) will identify this asterism. A 4 inch

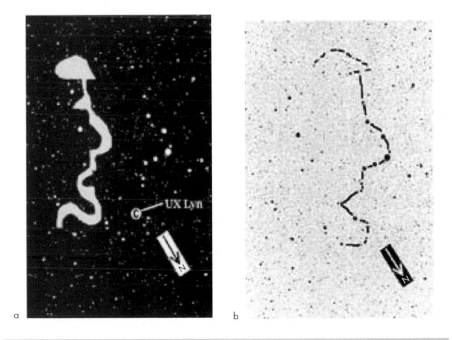

a   b

Figure 2.43. a °Note: Variable star. **b** Outline Face in Lynx ($1\frac{1}{2}$ deg. in size, 9 hr. 05 min. +38 deg. 20 min.).

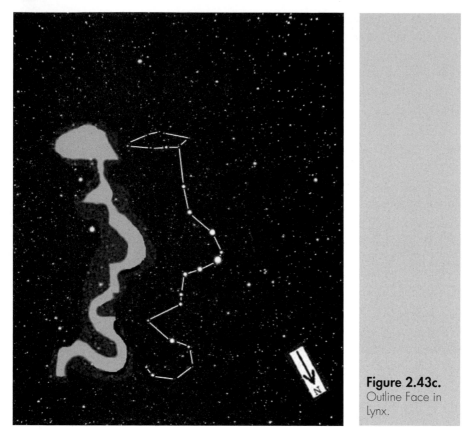

Figure 2.43c.
Outline Face in
Lynx.

(102 mm) wide field refractor at low power will bring you face to face with this $1\frac{1}{2}$ degree shape. At first glance I imaged this asterism as an outline face, although I don't think I had the pleasure of meeting this young lad before. He looks like a cashier at the local food store inquiring if I would like paper or plastic. The Sky Atlas shows four stars of the nose on the Outline Face asterism.

There are four 8 magnitude stars and brighter rounding the nose, with the brightest star being 5 magnitude and only 3 degrees northwest of star 38 Lynx. The Outline Face takes shape around 11 magnitudes when all of his stellar features are visible. This asterism is visible on the Mega Star 5 planetarium program at 11.5 magnitudes. The variable star UX Lyn, the lower lip of the Outline Face, does not vary enough to be noticeable in small telescopes. The small change in brightness that does occur will not interfere with his boyish smile. The fifth magnitude star on the nose of the Outline Face should have gotten a Flamsteed number because, after all, it is as plain as the nose on his face. It does take some time to get to know this pattern asterism, but take the time because a familiar face in a crowd of stars is always a welcomed sight.

Ursa Major Map and Asterisms

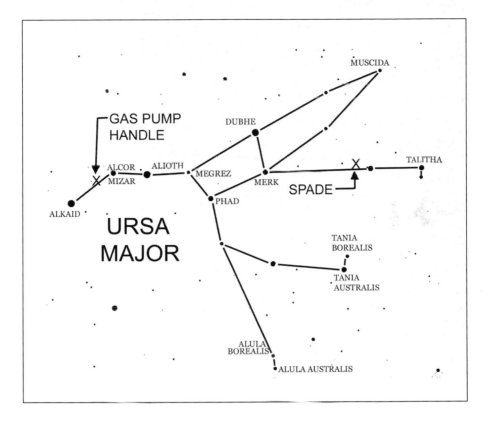

Spade

09 hr. 42 min.   +53 deg. 20 min.

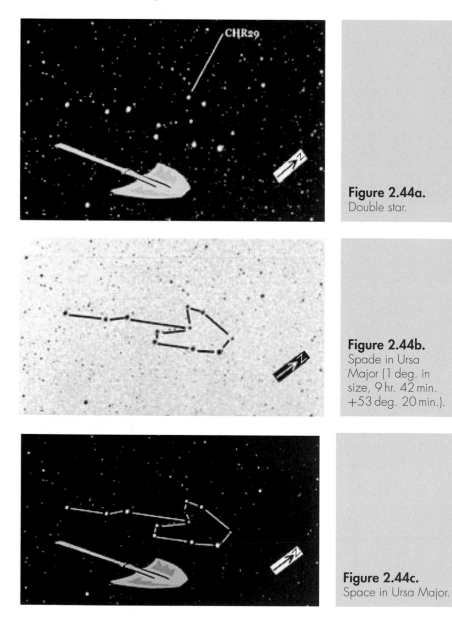

Figure 2.44a.
Double star.

Figure 2.44b.
Spade in Ursa Major (1 deg. in size, 9 hr. 42 min. +53 deg. 20 min.).

Figure 2.44c.
Space in Ursa Major.

It's time for a little planetary excavation with a stellar shovel. Like most intensive labor, this pattern asterism requires some optical muscle to see. A 50 millimeter binocular at 10 powers can only view a wavy line of stars. Eighty millimeter binoculars on a mount at 16 powers dig out this shape at around one degree in size. Some people commented that the Spade looks like a kite with a tail. Visualizing a reason to get out of work, Mmmmm? A 9.5 magnitude group of eight stars begins to outline the Spade, Fig. 2.44, with a few more stars at 10.5 magnitudes completing this shape. This is an easy pattern asterism with a few possibilities other then a work implement. A medium size telescope with a low F ratio should bring a few good suggestions at any star party.

The Spade is located between stars 26 and 30 Ursa Major and is not visible on the Sky Atlas 2000. There are nine stars visible on the Mega Star 5 planetarium program at 10 magnitudes. There is a tight double star, CHR29, located on the Spade right where a good foot push is needed for that healthy scoop of planetary dirt. There are 11 stars in this pattern asterism. During a presentation at Discovery Park for a group of third grades, I was told that the Spade was actually a stingray fish. Now that young lady has one clever imagination. So, if someone says it looks like a stingray in a clear pool of water, we have a pattern asterism in spades.

Gas Pump Handle

13 hr. 38 min.   +52 deg. 50 min.

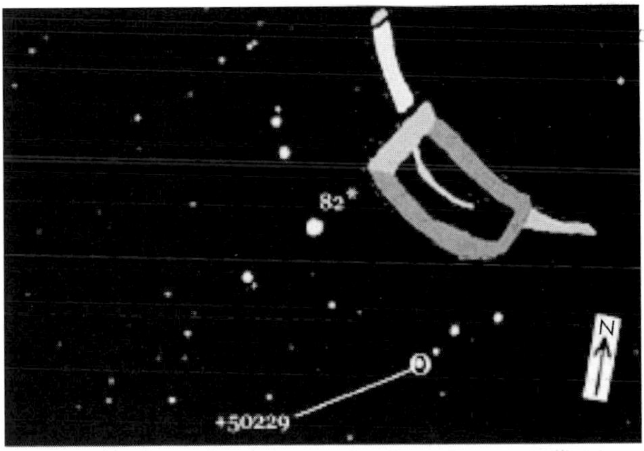

Figure 2.45a.
•Note: Flamsteed numbering system.
°Note: Variable star.

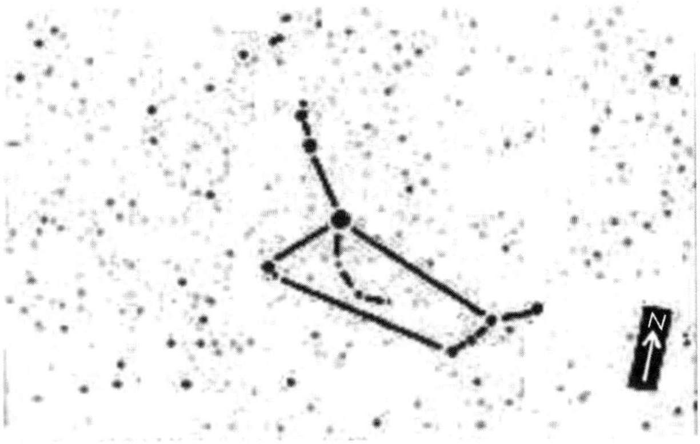

Figure 2.45b.
Gas Pump Handle in Ursa Major ($\frac{3}{4}$ deg. in size, 13 hr. 38 min. +52 deg. 50 min.).

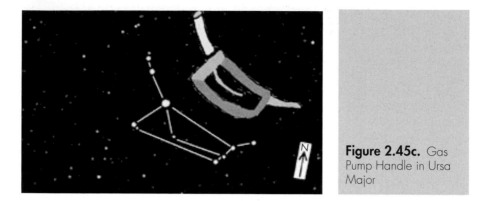

Figure 2.45c. Gas Pump Handle in Ursa Major

This pattern asterism is so obvious that it is in plain sight on the Sky Atlas 2000. At three-quarters of a degree in size, it fits in the field of the eyepiece just like a gas pump nozzle fits the opening of your gas tank. Five stars take on the basic shape at 7.5 magnitude and four more stars complete this pattern asterism at 9.5 magnitude. A 50 millimeter binocular at 14 power supplies just what is needed for a fill up at you local five star service center. This is a high octane party asterism for those states where you pump your own tank. The Gas Pump Handle, Fig. 2.45, is located in the handle (of course) of the Big Dipper between Mizar/Alcor and Alkaid.

The brightest star in this pattern asterism is six magnitudes, bright enough to have 82 Ursa Major as its own number. Besides the price of gasoline fluctuating, variable star +50229 in the handle of the gas pump has its own ups and downs in brightness. This star is seven magnitudes and may vary as much as $1\frac{1}{4}$ magnitude. Just as consumers check different gas stations for the most favorable gas prices, star +50229 should be closely watched for maximum and minimum luminosity. This variable star will not in any way affect the price of oil or diminish the Gas Pump Handle asterism shape. Any time you're checking out the horse and rider, be sure not to ignore this pattern asterism. Check the imaginative stellar sign floating off center in the twilight zone that says, last Gas Pump Handle stop until the Andromeda Galaxy!

Sextans Map and Asterism

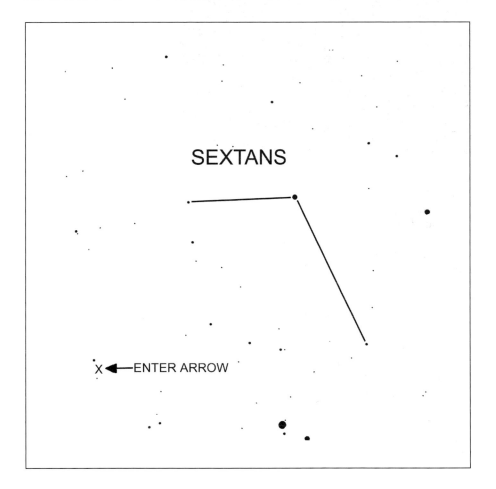

SEXTANS

X ←——ENTER ARROW

Enter Arrow

10 hr. 50 min.   −09 deg. 00 min.

Figure 2.46a.
•Note: Flamsteed
numbering system,
Double stars.

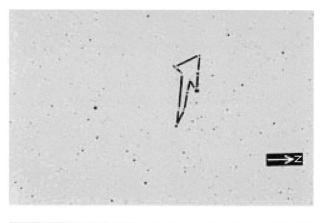

Figure 2.46b.
Enter Arrow in
Sextans ($\frac{3}{4}$ deg. in
size, 10 hr. 50 min.
−9 deg. 00 min.).

Figure 2.46c.
Enter Arrow in
Sextans.

Arrows point us in the proper direction and there are a million of them at airports. The trouble with arrows is that there's quite a few of them in the night sky. I only picked out two for this catalog and both can be seen in any busy city. This particular arrow asterism, Fig. 2.46, is telling folks which door to go in. After all, we don't want to knock a table waiter down with a full tray of Twinkies while going in the out door. The Enter Arrow is three-quarters of a degree in size and has two Flamsteed numbers attached to it. The brightest star in this asterism is six magnitudes and has number 41 Sextans assigned to it. The last star at the end of the Arrow's shaft is 39 Sextans. This pattern sits right on the border of Crater, and with Leo being just to the east. I have no doubt that this particular arrow points to the Lion's Den.

The eight other stars in this pattern asterism reach 10 magnitudes and require at least a 60 millimeter binocular at a steady 16 power to see. The Sky Atlas shows six stars with the two Flamsteed numbers being very apparent. Eight stars are visible on the Mega Star 5 at 10 magnitudes. Because of its size, telescopes would be a better choice. There are two double stars in this asterism. HJ838 should be easy to split with medium telescopes at one-quarter minute in separation. BU111 is a multi-double star of equal magnitude. The constellation of Sextons is in itself a location device, so finding an arrow in this instrument is totally expected. The Enter Arrow is not that difficult to see, but if you don't look hard enough, you might be going in the wrong direction.

Virgo Map and Asterisms

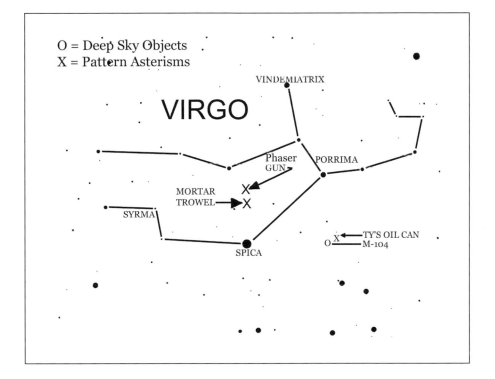

O = Deep Sky Objects
X = Pattern Asterisms

VIRGO

VINDEMIATRIX

Phaser GUN

PORRIMA

MORTAR
TROWEL

X
X

SYRMA

SPICA

TY'S OIL CAN
M-104

Ty's Oil Can

12 hr. 39 min.   −11 deg. 25 min.

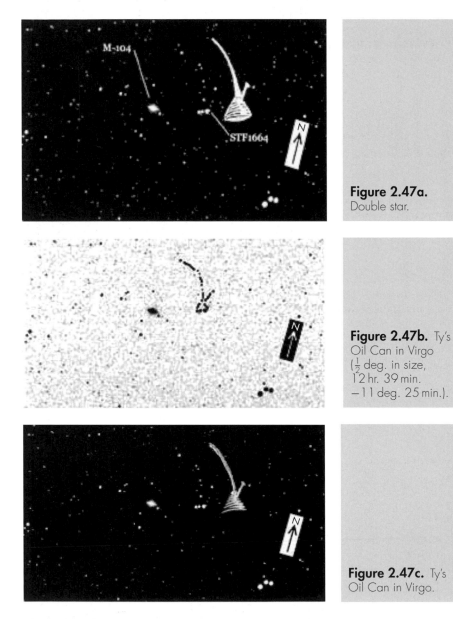

Figure 2.47a.
Double star.

Figure 2.47b. Ty's Oil Can in Virgo ($\frac{1}{2}$ deg. in size, 12 hr. 39 min. −11 deg. 25 min.).

Figure 2.47c. Ty's Oil Can in Virgo.

This is not a binocular pattern asterism unless you have those big 5 or 6 inch (127 or 152 mm) battleship coke bottles. Find someone with a medium to large telescope to zero in on this greasy shape because this is one oily can, Fig. 2.47. As you can tell from the picture, we have a drop of oil (M 104) from Ty's Oil Can and yes, HAZ-MAT has been notified! This pattern asterism is one-half degree in size and fits the description of a long necked, old fashion, button pump oil can. The three brightest stars at 9 and 9.5 magnitudes form the base of the oil can. The seven remaining stars form the pump and long neck of the oil can and they are a faint 11.5 magnitude each. The shape of this oil can is easy in any 6 to 10 inch (152 to 254 mm) reflector and the Sombrero Galaxy makes a cameo appearance as the oil spill.

To locate this oil can, squirt $11\frac{1}{2}$ degrees west of Spica. There is a very interesting multi-double star group right on the base right-hand side of the can. Star STF1664 can just be seen in the photograph at 12 magnitudes. The Mega Star 5 planetarium shows Ty's Oil Can perfectly with 10 stars at the 12 magnitude level. Only one star is visible on the Sky Atlas 2000, the multi-double star at 9 magnitudes, mentioned above. I am surprised that this little asterism has not been mentioned being so close to a popular galaxy. I have also surprised my son, Ty, by naming this oil can after him. It was only natural because Ty is one great auto technician. In his honor, this is Ty's Oil Can in the sky and that, I must conclude, greases the fittings for this pattern asterism.

Phaser Gun

13 hr. 20 min.   +03 deg. 00 min.

"Set your phasers on stun!" Captain Kirk ordered to his landing crew. "We don't want to injure any of the viewers. Oh! I mean natives of this planet." I remember that well as the little plastic model flew across the television screen. So, for old time sake, I decided to save this asterism instead of phasing it out. This Phaser Gun, Fig. 2.48, is bright in 50 millimeter binoculars at 10 powers. It is also visible in the Sky Atlas 2000, 2 degrees south of star 60 Virgo. Seven stars to 9.5 magnitude plus four stars at 10 magnitudes complete this pattern asterism. The Phaser Gun sizes out at $1\frac{1}{4}$ degrees.

The Phaser Gun has two double stars which have tight tolerances, totally expected from a futuristic blaster. The single variable star is located about where the power

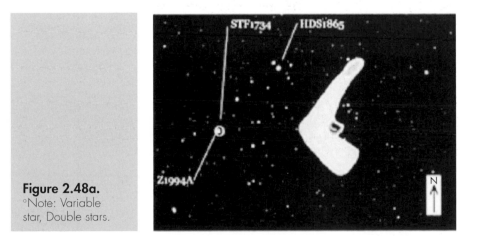

Figure 2.48a.
°Note: Variable star, Double stars.

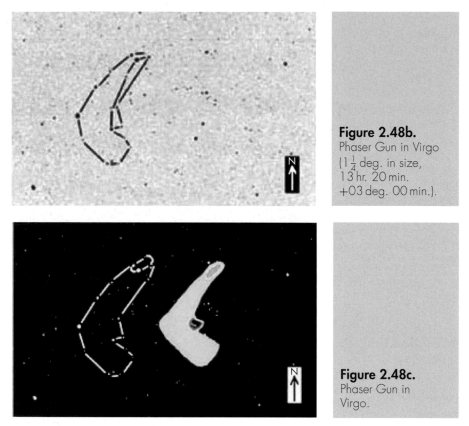

Figure 2.48b.
Phaser Gun in Virgo
($1\frac{1}{4}$ deg. in size,
13 hr. 20 min.
+03 deg. 00 min.).

Figure 2.48c.
Phaser Gun in
Virgo.

setting would be for a mild stun to a total cellular disruption. I must admit that it will take some imagination to trace out this pattern asterism. But, if you don't question warp drive and tribbles, the Phaser Gun should be within your orbital velocity.

Mortar Trowel

13 hr. 24 min.   −04 deg. 50 min.

Six stars show the general shape of the Mortar Trowel, Fig. 2.49, at 8.5 magnitude and brighter. Four more stars to 9.5 magnitudes cements this pattern asterism together. The triangle shape stands out immediately with a few stars above, right in line with the triangle, representing the handle. This shape is clearly visible in a 50 millimeter finder at 10 powers. The Mortar Trowel is one degree in size making it an easy target for binoculars and wide field telescopes alike. However, if you do your smoothing on clothes instead of between blocks, you might want to call this asterism an old fashion iron. If that doesn't get the wrinkles out, maybe a fancy cake knife will add a few calories to your imagination.

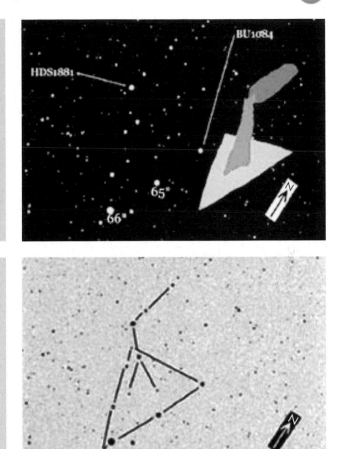

Figure 2.49a.
•Note: Flamsteed
numbering system,
Double stars.

Figure 2.49b.
Motar Trowel in
Virgo (1 deg. in
size, 13 hr.
24 min. −04 deg.
50 min.).

Figure 2.49c.
Motar Towel in
Virgo.

The Mega Star 5 planetarium program easily shows the Mortar Trowel pattern at 11 magnitudes. This is a bright pattern asterism worth the time to show and guess at star parties. Two double stars patch this shape together tightly, but BU1084 does have a separation of $2\frac{1}{2}$ seconds of an arc, just enough for that needed thermal expansion. For quick reference, the Mortar Trowel is only 6 degrees north of the bright star Spica in Virgo. The two bright stars in the Mortar Trowel are stars 65 and 66 Virgo and the triangle outline are clearly displayed on the Sky Atlas 2000.

Coma Berenices Map and Asterisms

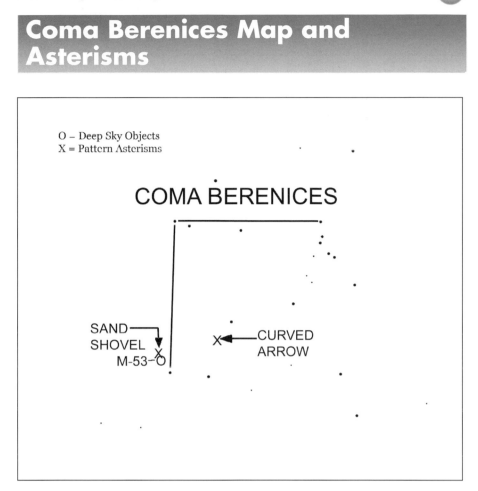

O – Deep Sky Objects
X = Pattern Asterisms

COMA BERENICES

SAND
SHOVEL
M-53
X—CURVED
ARROW

Curved Arrow

12 hr. 53 min.   +19 deg. 30 min.

I'm not meaning to throw you another curve, but this arrow is telling you to bend a little if you want to get where you're going. Nine stars show the curves well at 9.5 magnitude and much brighter. The tenth star at ten magnitudes adds the final bend at the end of the handle. This bright pattern asterism is easy in 50 millimeters binoculars at ten powers. It sounds like we are doing the power of tens here. Three quarters of a degree in size makes this little bent arrow accessible in most scopes at low power. The Curved Arrow, Fig. 2.50, is located about $1\frac{1}{2}$ degrees south of star 35 Coma Berenices or a little over 2 degrees south from that faint fuzzy M 64. In fact, this helpful arrow is pointing toward an area in Coma Berenices where there are loads of faint fuzzes. I wonder who maintains these directional stellar buoys of the heavens.

The Curved Arrow is very obvious in the Mega Star 5 planetarium program at 9.5 magnitudes. The Sky Atlas 2000 plots seven stars for location. Double star STF 1685

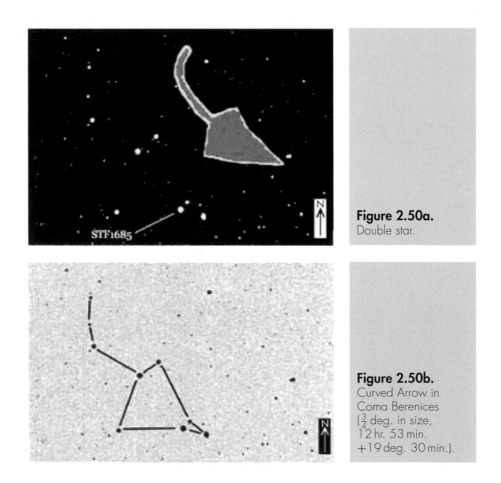

Figure 2.50a.
Double star.

Figure 2.50b.
Curved Arrow in Coma Berenices ($\frac{3}{4}$ deg. in size, 12 hr. 53 min. +19 deg. 30 min.).

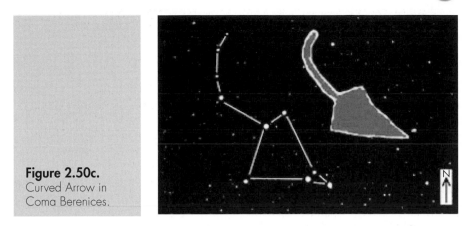

Figure 2.50c.
Curved Arrow in Coma Berenices.

has two stars at 7.3 magnitudes and 7.8 magnitudes with a split of 16 seconds between them. The Curved Arrow is the second and last arrow in this catalog. After all, I have to be careful; I'll get a reputation for pointing too much! This asterism has a bright 6 magnitude star involved in its pattern and should have deserved a numeral. However, with the Coma Cluster looking like a numbered dot to dot drawing waiting to be outlined, this bright star in the arrow was certainly overlooked for a Flamsteed notation. Just goes to show that having an important job like pointing out galactic neighbors doesn't mean that there's compensation to be had.

Sand Shovel

13 hr. 15 min.   +19 deg. 00 min.

We have often heard of stars in our galaxy being represented as grains of sand. Needless to say, I have found the very Sand Shovel, Fig. 2.51, used in that calculation. This

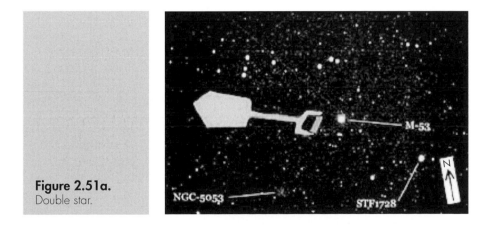

Figure 2.51a.
Double star.

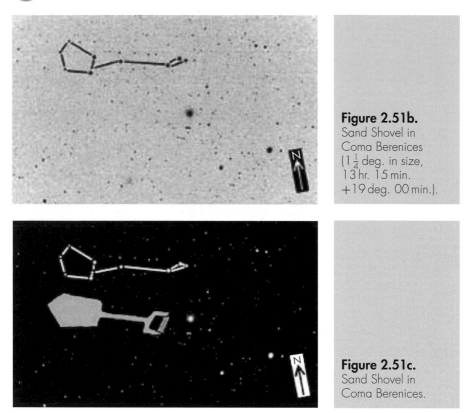

Figure 2.51b.
Sand Shovel in Coma Berenices ($1\frac{1}{4}$ deg. in size, 13 hr. 15 min. +19 deg. 00 min.).

Figure 2.51c.
Sand Shovel in Coma Berenices.

asterism is now busy shoveling star dust from the grand globular cluster M 53 into the Berenices star field, setting it aglow. The Sand Shovel is completely visible in the Sky Atlas 2000 just above M 53 and NGC 5053, another globular cluster. I have introduced this shape to many people and nobody has had trouble seeing the shovel. The Sand Shovel is $1\frac{1}{4}$ degrees in size and can fill a sea shore play pail with thousands of sand grains/stars at the speed of light. This pattern asterism is an excellent showpiece in 50 millimeter binoculars at 10 powers and a definite must at star parties. Ten stars make up the Sand Shovel at 9 magnitude and brighter. 4 and 5 inch (102 and 127 mm) wide field refractors will do justice to both the globular clusters and the shovel very close by.

I have pointed out a double star for reference. The double star is not involved in the Sand Shovel asterism, but with the few faint stars by it, we could put together a little sand shell as a promotional prop with no problem. The Sand Shovel is also visible on the Mega Star 5 planetarium program at 9 magnitudes. This pattern asterism employs the deep sky objects next to it (the two globular clusters) as a theme. The shovel is only 5 degrees east of the Curved Arrow. The bright multi-double star STF1728, also known as Flamsteed number 42, is only $1\frac{1}{2}$ degrees south of the Sand Shovel. Living entirely too far from the beach should be a good reason to scoop up the Sand Shovel next time you're outside at night observing the ocean of stars above.

Scorpius Map and Asterisms

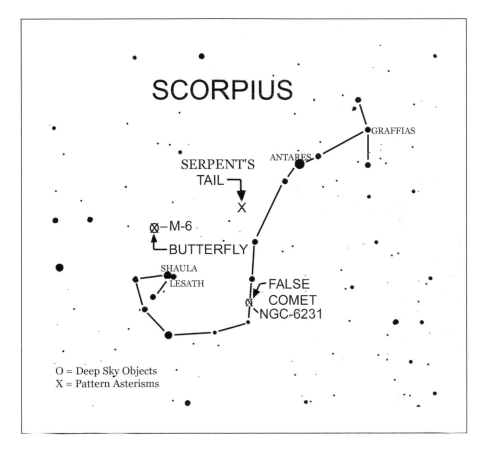

False Comet

16 hr. 53 min.   −41 deg. 30 min.

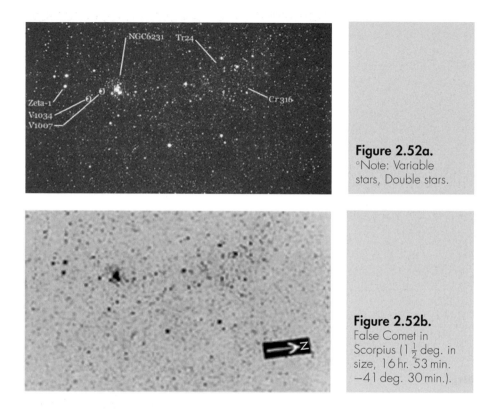

Figure 2.52a.
°Note: Variable stars, Double stars.

Figure 2.52b.
False Comet in Scorpius (1$\frac{1}{2}$ deg. in size, 16 hr. 53 min. −41 deg. 30 min.).

These splashes of stars are a well-known astronomical handful. The further south and the less humidity you have, the better they can be seen. The False Comet, Fig. 2.52, has been identified as an interesting eyeball object not requiring optical aid for thousands of years. Thirty millimeter binoculars at 6 powers will bring out the cluster of stars forming the head of the comet. It will also show the stream of stars making up the tail of the False Comet. If you are new to astronomy with only binoculars at your side, this is a top notch go getter. This open cluster/asterism mix is 1$\frac{1}{2}$ degrees in size and contains three open clusters.

The head of the False Comet is NGC 6231, a compact and most beautiful open cluster. Star Zeta-1, a half of a degree south from this cluster, is thought to be an outlying cluster member. There are no bright stars in this asterism mix, but the total stellar combination makes the False Comet bright and obvious. Along with NGC 6231, open cluster Collinder 316 and Trumpler 24 make up the entire False Comet. To locate this stationary fireball, find stars 35 and 34 Scorpius. That is the end of the tail where the stinger is. Follow the five stars west around the tail from the stinger and you are there. This area is littered with double stars and variable stars making this region of space variably doubled in stellar curiosities.

Serpent's Tail

16 hr. 55 min.   −31 deg. 00 min.

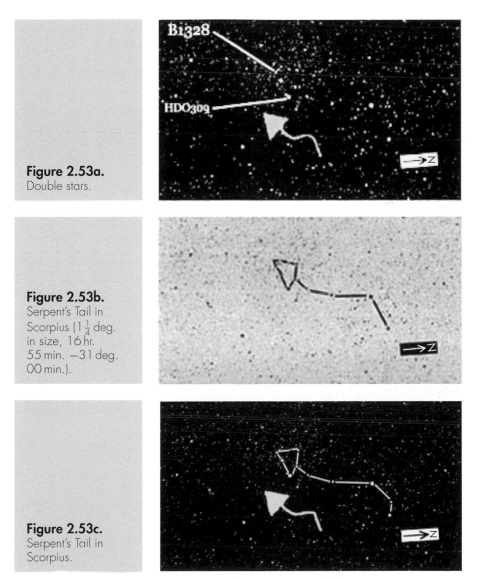

Figure 2.53a.
Double stars.

Figure 2.53b.
Serpent's Tail in Scorpius (1 $\frac{1}{4}$ deg. in size, 16 hr. 55 min. −31 deg. 00 min.).

Figure 2.53c.
Serpent's Tail in Scorpius.

A pointy long tail we normally associate with being attached to a dragon, serpent or devil, this pattern asterism is right at home riding on the back of Scorpius the Scorpion which possesses a formidable tail itself. Ten stars shape out the Serpent's Tail, Fig. 2.53, at 8.5 magnitude. Eight more stars come into play one magnitude fainter to complete this back end of a lot of trouble. This group of stars can be seen in 50 millimeter binoculars at 10 powers and it covers a field of view 1 $\frac{1}{4}$ degrees in size. Most of the tail is visible in the Sky Atlas 2000, only $1\frac{1}{2}$ degrees west of M 62, a globular cluster. This is another bright pattern asterism located in an area of wonderful stellar showpieces.

There are no fewer then six double stars in this pattern. I guess you can call this an asterism a lot of double trouble. B1328 is a multi-double and as one should expect, located very near the tip of the Serpent's Tail. The Mega Star 5 planetarium program reveals 12 stars from the Serpent's Tail perfectly at 8.5 magnitudes. It has been suggested to me by one viewer that this shape looks like a ladle for a punch bowl. I can see that, perhaps more clearly if the ladle had a little extra punch to it in the bowl. But in this case, I kept my long ago college party thoughts to myself. You can view this pattern asterism with safety knowing that real trouble is, since we are viewing the back end, moving away from us.

Butterfly

17 hr. 40 min.   −32 deg. 13 min.

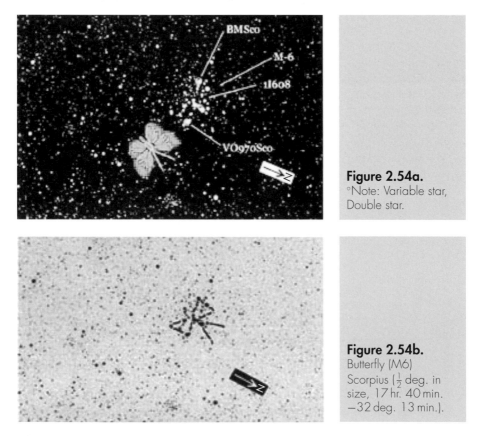

Figure 2.54a.
°Note: Variable star, Double star.

Figure 2.54b.
Butterfly (M6) Scorpius ($\frac{1}{2}$ deg. in size, 17 hr. 40 min. −32 deg. 13 min.).

The Butterfly, Fig. 2.54, open cluster, M 6, is a well-known deep sky object to all observational astronomers everywhere. It is a half-degree in size with stars from 6 to 10 magnitudes. Fifty millimeter binoculars with 16–20 powers on a steady mount will catch this little fellow in flight. A 60 millimeter spotting scope at 20 powers will show

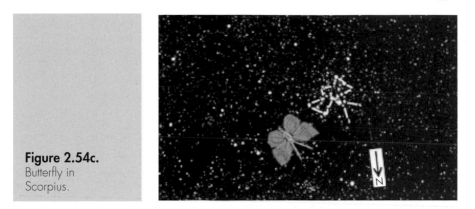

Figure 2.54c.
Butterfly in
Scorpius.

the field well. Telescopes up to 6 inches (152 mm) at low power will show many stars, including the two antennae most evident of a butterfly. M 6 is 3 degrees northwest of M 7, a large open cluster visible by eye with dark skies. The entire area is a stellar wonderland of our Milky Way Galaxy and is a treat with most any scope.

There are a few double stars in M 6 and I noted one, 1I608, a tight double for reference. There are more variable stars then double stars in this cluster and I mentioned two of interest. VO862 and BM Sco are two of the brightest stars in M 6 and they both vary noticeably. These stars are located on the right and left wings of the butterfly. Although these two variables will not change the shape of the butterfly, time lapse photography may give the appearance of wing movement. BM Sco has a period of $2\frac{1}{4}$ years, so if you have several pictures of M 6 taken over many years, it's time to compare. If you are just starting out in astronomy, flutter on over to the Butterfly and be sure to give this pattern open cluster a few drops of optical nectar.

Draco Map and Asterism

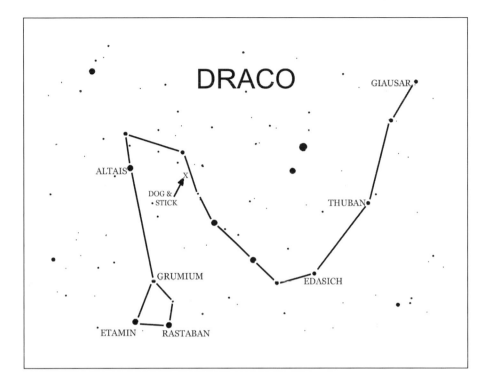

Dog and Stick

17 hr. 36 min.   +68 deg. 40 min.

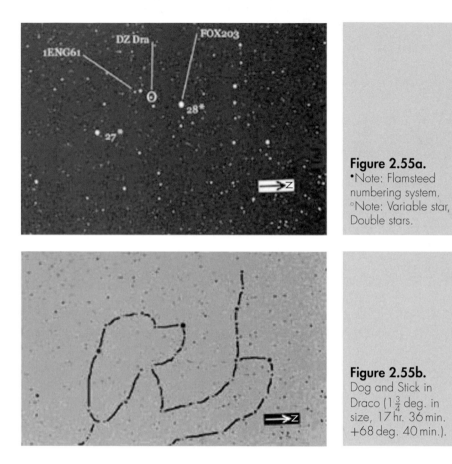

Figure 2.55a.
•Note: Flamsteed
numbering system.
°Note: Variable star,
Double stars.

Figure 2.55b.
Dog and Stick in
Draco (1$\frac{3}{4}$ deg. in
size, 17 hr. 36 min.
+68 deg. 40 min.).

This pattern asterism is absolutely remarkable. Most of the outline is visible in a 50 millimeter finder scope at 10 powers. Sixty millimeter binoculars at a steady 18 power bring out the complete shape. This asterism is fantastic in 4 or 5 inch (102 or 127 mm) wide field refractors at low power. The only symbols missing from this dot to dot pattern are the numbers next to the dots. This bright pattern asterism is 1$\frac{3}{4}$ degrees in size. Stars 27 and 28 Draco, 5.5 magnitudes and 6.0 magnitudes, are the bright nose and back of the head of the dog. The lower jaw and forehead of the dog is most conspicuously outlined. The lined negative photo will help you fill in the remaining shape. The Dog and Stick, Fig. 2.55, stand out well among the background stars. This pattern will delight many onlookers as they too trace out the figure in the eyepiece.

Most of the pattern is composed of 8.5–10 magnitude stars and 11 stars are visible on the Sky Atlas 2000. On the Mega Star 5 planetarium program, the Dog and Stick could not look more obvious as this stellar puppy looks northward in the sky. There is one variable star on the dog's forehead that will not change the way you see this pattern. A double star of interest is 1ENG61 which has an 8 magnitude and a 9 magnitude star at an average of $1\frac{1}{2}$ minutes separation. The Dog and Stick is only $3\frac{1}{2}$ degrees south of star 31 Draco. So check out this little buddy in the sky because he gets to go on vacation with the family to that dark viewing sight next time around, no doubt about it!

Hercules Map and Asterisms

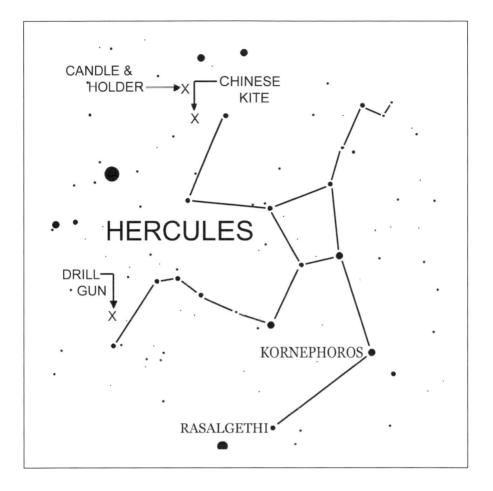

Chinese Kite

17 hr. 58 min.　　+45 deg. 50 min.

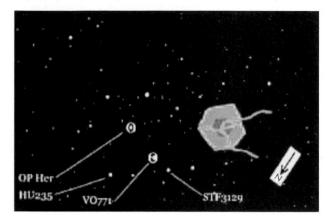

Figure 2.56a.
°Note: Variable
stars, Double stars.

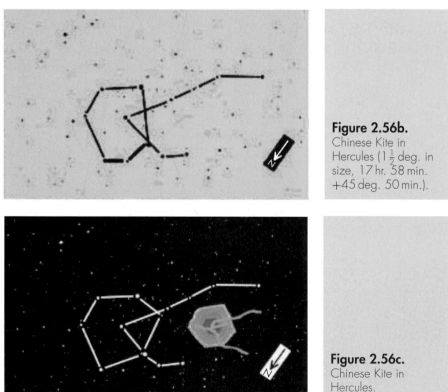

Figure 2.56b.
Chinese Kite in
Hercules (1½ deg. in
size, 17 hr. 58 min.
+45 deg. 50 min.).

Figure 2.56c.
Chinese Kite in
Hercules.

Flying high over Hercules is a little Oriental paperwork on chopsticks. The Chinese Kite, Fig. 2.56, takes flight with a basic shape of nine stars, filled in by four, 9.5 magnitude kite strings. This asterism stands out well from the background stars and is 1½ degrees in size. A 50 millimeter binocular at 16 steadied power displays this six sided high flyer nicely. The Chinese Kite is in the sky about 3 degrees east of star 85 Hercules. Four of the brighter stars of the kite are 8 magnitude and show up on the Sky Atlas 2000. A few people viewing the Chinese Kite through the eyepiece of my 5 inch (127 mm) refracting telescope agreed with me and one did not. This particular person saw the Star Ship Enterprise leaving the atmosphere after going back in time to Earth. No doubt a Trekkie from the old series, but I have seen that particular episode also and nodded my head in approval.

The Chinese Kite is most visible on the Mega Star 5 planetarium program at 8.5 magnitude. If you use more magnitude on this planetarium program, the kite will become lost in the stars. I made note of two double stars on the photo. STF3129 has a 7.5 magnitude star and an 11.0 magnitude star separated by one-half minute. That double star should not be difficult to split no matter what atmosphere you go back in time to. Of the two variable stars, VO771 only dims 0.8 magnitudes and that would not cause any asterism shape shifting. The Chinese Kite should be a high flying attraction at star parties where the guessing will turn into fireworks. And like all of the *kites* I have sailed into the blue, remember to sight this pattern asterism before it gets too low in the sky where it may get lost behind the trees.

Candle and Holder

18 hr. 03 min.   +48 deg. 00 min.

This is a very interesting and obvious pattern asterism Hercules used to find his way in the dark. Maybe I should have made it an Olympic torch; I don't think candles were invented just yet. Oh well, light this shape up in your 60 millimeter binoculars at 16 powers on a mount and you will get a good view of this area in the dark. Six bright stars in the Sky Atlas 2000 display this group and it appears on the map as a sword. This guiding light is $1\frac{1}{4}$ degrees in size and in the positive photo the Candle and Holder, Fig. 2.57, look more like a boat anchor then anything, but it is too late to change the name now. I'll leave that up to the observing individuals. This asterism is quite evident on the Mega Star 5 planetarium program at 10.5 magnitude. Since it

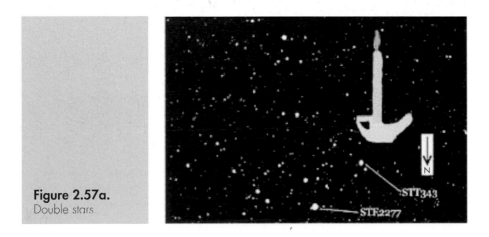

Figure 2.57a.
Double stars

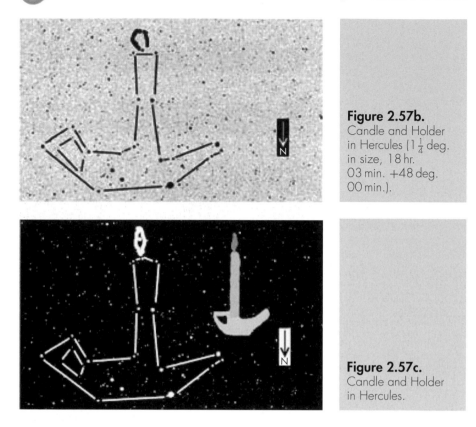

Figure 2.57b.
Candle and Holder in Hercules (1$\frac{1}{4}$ deg. in size, 18 hr. 03 min. +48 deg. 00 min.).

Figure 2.57c.
Candle and Holder in Hercules.

is upside down it appears to look more like a crossbow. Gee whiz, maybe we should develop a Swiss Army asterism with all of these handy attachments?

The Candle and Holder is 2 degrees east of the star 88 Hercules and the outline is visible in a 50 millimeter finder scope at 10 power. The Candle and Holder is also 2 degrees above, or north, of the Chinese Kite, Fig. 7.2. Five stars of 8 magnitudes form the basic shape of the Candle and Holder. Nineteen stars finish the pattern asterism at 9.5 magnitude. STT343 is a double star with a 7.5 magnitude star and a 10.5 magnitude star with a separation of about thirteen seconds of an arc. The other double, STF2277, is more interesting, being a multi-double star system. Lots of double A's powering this wireless light. This asterism should hold second light to the very bright M 13 and M 92, both outstanding globular clusters and that is a lot of candle power!

Drill Gun

18 hr. 26 min.   +26 deg. 10 min.

Thank goodness for those battery operated, hand-held, go anywhere, mechanical termites, those friendly, two speed, reversible, rechargeable munch masters of the woodworking shop. When I found this three-quarter of a degree work implement in

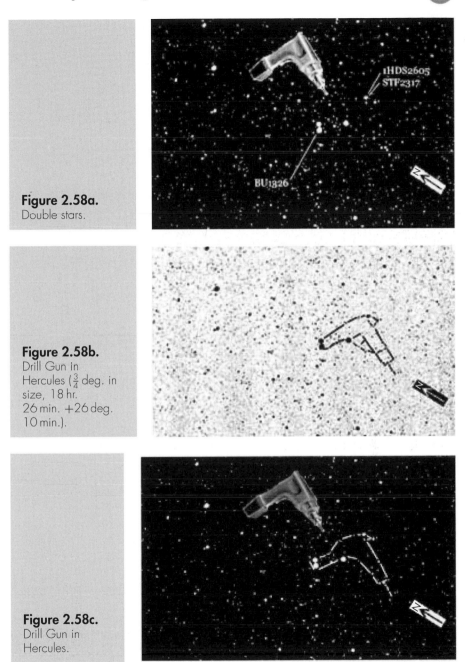

Figure 2.58a.
Double stars.

Figure 2.58b.
Drill Gun in
Hercules ($\frac{3}{4}$ deg. in
size, 18 hr.
26 min. +26 deg.
10 min.).

Figure 2.58c.
Drill Gun in
Hercules.

my eyepiece, I knew my job had just gotten a lot easier. Despite all of this fanfare, this pattern asterism is not that easy to see at first. Sixty millimeter binoculars at 16–20 powers on a mount for steady drilling will carve out the Drill Gun, Fig. 2.58 in the sky. Small to medium telescopes at low power will provide lots of needed amperage for a clean view. The Drill Gun is abounding with stars that may hinder its appearance, but once you see it in the sky, you will always spot it right off the next time around. The Drill Gun is $2\frac{1}{2}$ degrees south of star 107 Hercules and is just on the boarder of Lyra.

Four 8 magnitude stars represent the back end of the Drill Gun. Two bright stars at the bottom of the handle symbolize the battery. Further up the handle is another bright star representing the trigger and the last star is the back of the Drill Gun. The remaining nine stars go to 10 magnitudes to complete this asterism. BU1326, a multi-double star, is located in the battery section of the Drill Gun. It has a 6.5 magnitude star and an 8.5 magnitude star with a tight separation of one minute of an arc. The back end of the motor housing also has a multi-double labeled STF2317. The Mega Star 5 planetarium program shows the Drill Gun hanging handle first on the stellar pegboard in the Herculean workshop in plain view at 10 magnitudes. Combine this pattern asterism fully charged with a piece of wood and its bon appetite.

Ophiuchus Map and Asterism

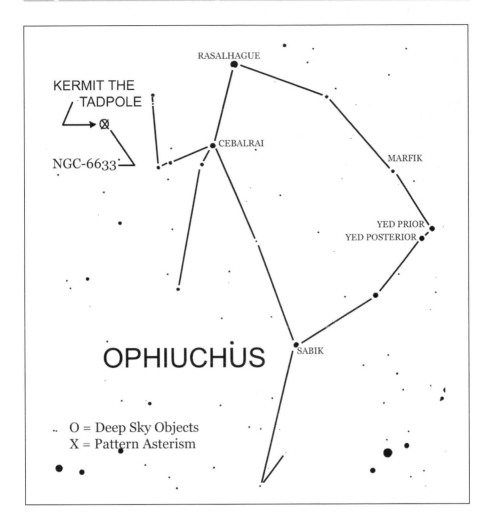

KERMIT THE
TADPOLE

RASALHAGUE

CEBALRAI

NGC-6633

MARFIK

YED PRIOR
YED POSTERIOR

OPHIUCHUS

SABIK

O = Deep Sky Objects
X = Pattern Asterism

Kermit the Tadpole

18 hr. 27 min.   +06 deg. 30 min.

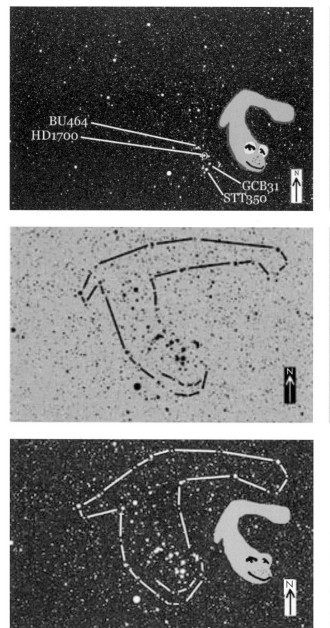

Figure 2.59a.
°Note: Variable star, Double stars.

Figure 2.59b.
Kermit the Tadpole in Ophiuchus ($\frac{3}{4}$ deg. in size, 18 hr. 27 min. +6 deg. 30 min.).

Figure 2.59c.
Kermit the Tadpole in Ophiuchus.

Now this is a picture that Ms. Piggy will have on her night stand. In this particular open cluster/asterism mix, the eyes tell the entire story. In the open cluster NGC 6633, there are two groupings of stars that slant Oriental style. This section of stars in the open cluster forms the face of baby Kermit. I'll bet you didn't know Kermit had freckles when he was a little wiggle in a pond? At three-quarters of a degree in size, Kermit the Tadpole, Fig. 2.59, must have been a tiny handful. Eighty millimeter binoculars at 20 powers on a mount will be needed to separate Kermit's eyes in this cluster, but NGC 6633 is visible as a faint swarm of stars in a 50 millimeter finder at 10 powers. This little tadpole sits on the boundary of Serpens Cauda, only 4 degrees north and a tiny east of star 74 Ophiuchus. The entire pattern becomes clear around 10 magnitudes. A medium size telescope at low power will really ham up this tadpole, or is that the nickname of his soon to be girlfriend?

I have found out from my asterism lectures that third graders are not familiar with the Muppets, so it is up to the earlier generations to appreciate this pattern mix. There is one variable star noted on the photo that will not change the shape of this pattern mix. Three double stars adorn the face of Kermit the Tadpole. A freckle of interest is GCB31 containing two 9.5 magnitude stars separated by $3\frac{1}{2}$ seconds of an arc. NGC 6633 is a beautiful open cluster and if it were not for the two groupings of stars making up Kermit's eyes, why then this pattern would not have a chance to grow legs and leave the pond.

Vulpecula Map and Asterism

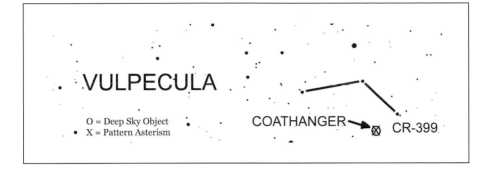

Coathanger

19 hr. 26 min.   +20 deg. 00 min.

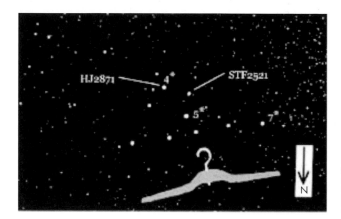

Figure 2.60a.
•Note: Flamsteed numbering system, Double stars.

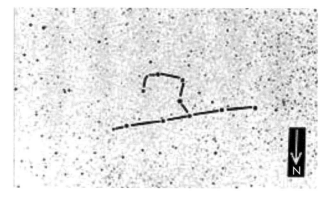

Figure 2.60b.
Coathanger in Vulpecula (1$\frac{1}{2}$ deg. in size, 19 hr. 26 min. +20 deg. 00 min.).

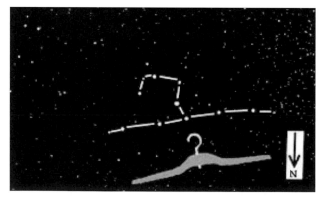

Figure 2.60c.
Coathanger in Vulpecula.

A most superior and well-known asterism, open cluster mix; Cr-399 is a great view in most any binoculars. A 50 millimeter finder scope at 10 powers will take this hanger with winter coat right to the cleaners. All 10 stars are 7.5 magnitudes or brighter and besides being called the Coathanger, Fig. 2.60, some see this shape as a top hat. Turn the image upside down and a playground seesaw might appear. Another suggestion put forth by one young lad was a sailboat. A pair of sharp eyes and a dark desert sky can see this group mixed in with the Milky Way. The Coathanger is $1\frac{1}{2}$ degrees in size and hangs on to three Flamsteed numbers of their own. Sometimes called Brocchi's Cluster, these stars may not all belong to the Collinder 399 cluster, so I have been informed through the clothesline, or was it the grapevine.

Two interesting doubles stars are hooked onto the hook of this hanger. HJ 2871 has a 5.2 magnitude star and a 10.0 magnitude star that are separated by an average 22 seconds of an arc. The second is STF2521 with a 5.8 magnitude star and a 10.5 magnitude star. They are separated by 24 seconds of an arc. The Coathanger is hanging upside down and in plain view on the Sky Atlas 2000. Cr-399 is only 8 degrees south of that very famous colorful double star Albireo, located in the constellation of Cygnus. This mix is a must at in the season star parties. So, when relaxing in your outside easy chair with a pair of binoculars and scanning the Milky Way, stop over at Vulpecula and hang ten.

Aquila Map and Asterism

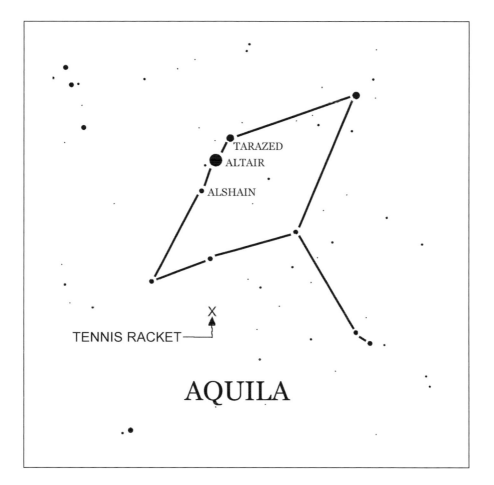

Tennis Racket

19 hr. 50 min.   −04 deg. 35 min.

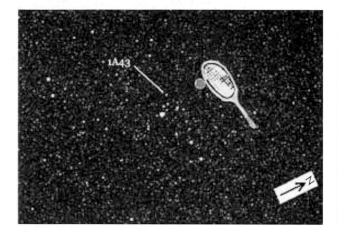

Figure 2.61a.
Double star.

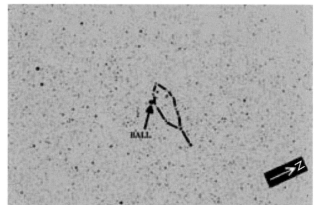

Figure 2.61b.
Tennis Racket in
Aquila ($\frac{3}{4}$ deg. in
size, 19 hr. 50 min.
−04 deg. 35 min.).

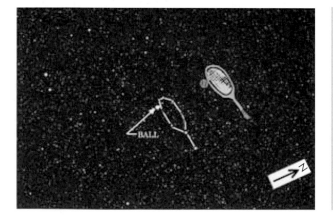

Figure 2.61c.
Tennis Racket in
Aquila.

For the tennis fans out there, on center court with tennis ball in the serve position, we have the Tennis Racket pattern, Fig. 2.61. Better be close to the field of play to get a good look at this action packed asterism. The most obvious star at 7 magnitudes is the tennis ball. The next two stars at 8.5 magnitude each is the handle of the Tennis Racket. The remaining six stars that make up the oval of the racket start at 9.5 magnitude and string out to 11 magnitudes. At least 60 millimeter binoculars at a steadied 16–20 powers will bring out the entire shape. This pattern is faint in a 50 millimeter finder scope at 10 powers. Located 3 degrees east of star 42 Aquila, the Tennis Racket has lots of Milky Way stars in the background that do not interfere with the game. At three-quarters of a degree in size, a medium size telescope of one's own liking will do justice to this asterism.

Only three stars appear on the Sky Atlas 2000 and they are the two stars for the racket handle and the tennis ball. I have only one double star noted for location, but 1A43, a tight pair, is only a spectator in the stadium and does not belong to the playing field racket. This tennis racket is very visible on the Mega Star 5 planetarium program at 9.5 magnitudes. Not too difficult to image, this kind of sports racket may just be your kind of play.

Cygnus Map and Asterisms

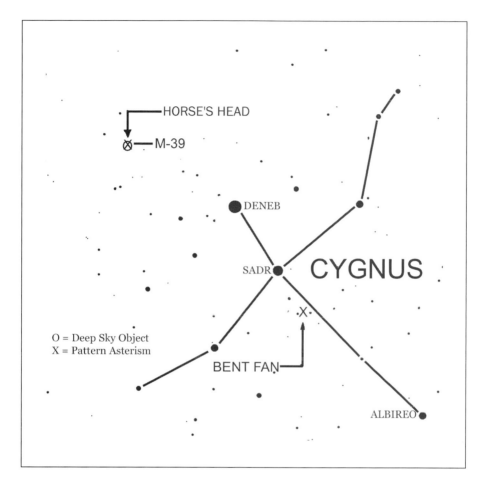

Bent Fan

20 hr. 14 min.   +36 deg. 35 min.

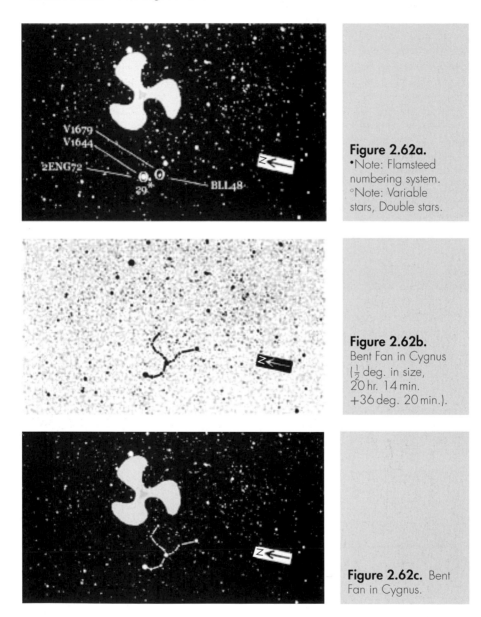

Figure 2.62a.
•Note: Flamsteed numbering system.
°Note: Variable stars, Double stars.

Figure 2.62b.
Bent Fan in Cygnus ($\frac{1}{2}$ deg. in size, 20 hr. 14 min. +36 deg. 20 min.).

Figure 2.62c. Bent Fan in Cygnus.

Buried deep in the Milky Way is a group of stars that look something close to a Klingon (Star Trek) Government symbol. In case you are not a Trekkie, the pattern asterism Bent Fan, Fig. 2.62, will do nicely. A mounted 80 millimeter binoculars at 16–20 power will do, but a telescope is suggested for this faint, one-half degree grouping of stars. Some say it looks like a buzz-saw blade in motion. With this interesting shape, imaginations can run rampant. Special attention was given not to over-expose the film when photographing this pattern asterism due to the Milky Way washout effect. The brightest star in the group is 29 Cygnus at 5.5 magnitude. There are 11 stars all together, most 9 or 10 magnitude.

The Sky Atlas 2000 shows four stars but no clue about the fan shape. Lots of bright stars in the area will make finding this asterism a hit of miss affair unless you have a programmable go-to mount. The Bent Fan is easy to visualize on the Mega Star 5 planetarium program at 9 magnitudes. Increase the magnitude to 10.5 and the shape disappears among the background stars. Star 29 Cygnus is both a variable star and a double star. As a double star, it has an 8 magnitude star and a 10 magnitude star separated by 2 minutes of an arc. The Bent Fan stands out well from all of the stars around it and is a teaser when playing guess the mess at star parties.

Horse's Head

21 hr. 32 min.   +48 deg. 30 min.

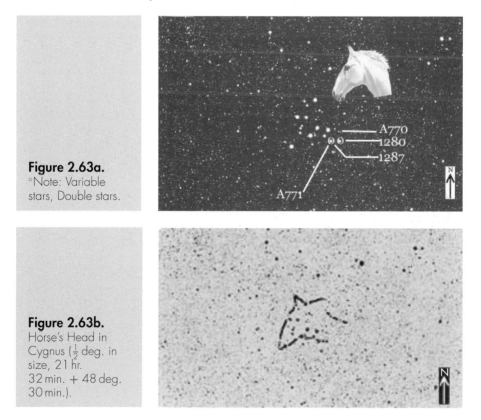

Figure 2.63a.
°Note: Variable stars, Double stars.

Figure 2.63b.
Horse's Head in Cygnus ($\frac{1}{2}$ deg. in size, 21 hr. 32 min. + 48 deg. 30 min.).

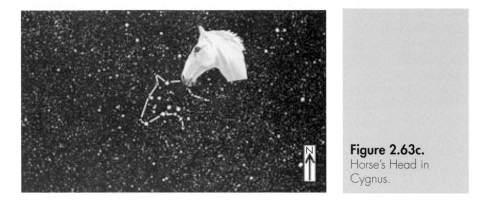

Figure 2.63c.
Horse's Head in Cygnus.

I love open clusters for their dot to dot appeal and M 39 when seen with low power is a good example. Using high power this cluster looks just like a splash of stars. At first I saw this open cluster as a bouquet of flowers or a feather duster. However, I have had several young people tell me during my presentations at Discovery Park in Safford, Arizona, that it looks just like a horse's head and a little part of the neck. What a wonderful imagination the young people have. I couldn't let that suggestion pass me by because I could see that shape also. The last bright star to the left of the cluster in the picture is the horse's nose. The northern-most star in the photo is the tip of the ears and the five stars bottom right on the photo are the neck. If you can see that image, you have just graduated pattern asterism room 101, just like I have.

M 39 is easy in 30–50 millimeter binoculars at 16 powers. All of the outlining stars form the Horse's head, Fig. 2.63. As is normal with open clusters, the outline stars shape the pattern and the remaining stars inside the pattern are fill. I have noted two double stars and two variable stars. 1287 and 2A771 belong to one star of dual nature. The double stars are tight and the variables vary very little. M 39 is visible in any small finder scope by following star 58 to star 62 and then $6\frac{1}{2}$ degrees in their general direction. If you present this open cluster to people that have allergies and sneeze around horses, well used feather dusters or flowers, then I suggest that you display M 39 as just a simple splash of stars.

Pegasus Map and Asterisms

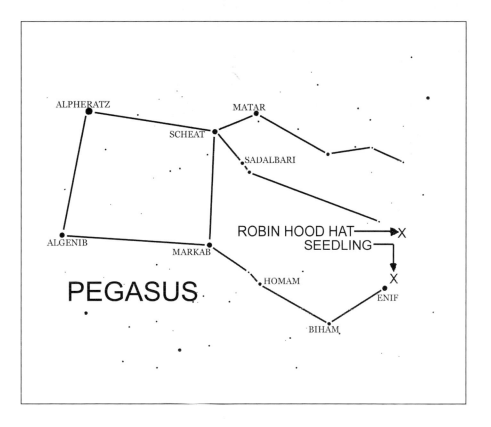

Robin Hood Hat

21 hr. 36 min.   +15 deg. 30 min.

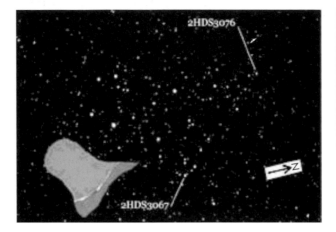

Figure 2.64a.
Double stars.

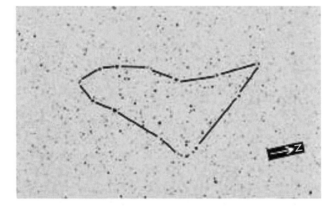

Figure 2.64b.
Robin Hood Hat in Pegasus (1 ½ deg. in size, 21 hr. 36 min. +15 deg. 30 min.).

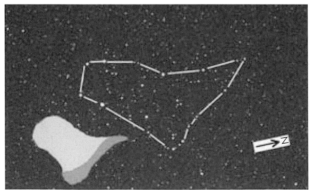

Figure 2.64c.
Robin Hood Hat in Pegasus.

Composed of bright and faint stars, this pattern asterism outline is visible in a 50 millimeter finder at 10 powers. Twelve stars to 9.5 magnitude form the shape of the Robin Hood Hat, Fig. 2.64. The very first setback about this Sherwood Forest mop cover is, where's the feather in the hat, asked by a third grade classroom student. My reply, the second time around, was that Mr. Hood either uses it for his arrows or the feather is on the other side of the hat, out of view. Unfortunately, the first time around, I was stumped like a tree in the forest. There are lots of stars in the hat, but nothing shaped like a feather. Another suggestion from one imaginative young person was that this group of stars looked like a cowboy boot. Well, that takes care of the feather problem but now there are no spurs for that get up and go. At that point I was itching to click to the next slide.

The Robin Hood Hat is located about 3 degrees southwest of star 9 Pegasus and the Sky Atlas 2000 shows seven stars outlining the top of the hat. The Mega Star 5 planetarium program details the outline of the hat very well at 9.5 magnitudes. Sixty millimeter binoculars at 14 powers will show the entire hat, minus the feather, in dark skies. There are two very tight double stars in this pattern asterism for location. Whether you see a cowboy boot, Sir Robin's hat or a beehive full of stars, this is an interesting shape. Using a wide field 4 inch (102 mm) refractor at low power will not only show you the hat, but the hat size, which in this case is about $1\frac{1}{2}$ degrees, or a Wal-Mart large.

The Seedling

21 hr. 43 min.   +10 deg. 20 min.

This is a very interesting but faint pattern asterism, so put away the binoculars. The shape only starts to appear around 10 magnitudes and is completed around 11.5–12 magnitude. At a little under three-quarters of a degree in size, The Seedling, Fig. 2.65, can be planted visually in a 6 inch (152 mm) reflector at low power. It is super-easy to find because it is only one-half degree from star 8 Pegasus. There are no stars visible in the Sky Atlas 2000. What's really nice is the globular star cluster M 15 is just up the

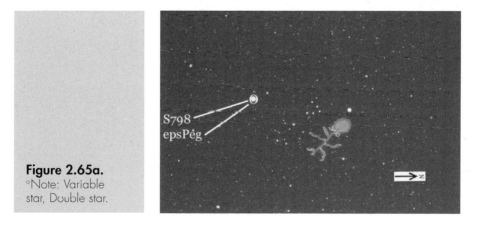

Figure 2.65a.
°Note: Variable star, Double star.

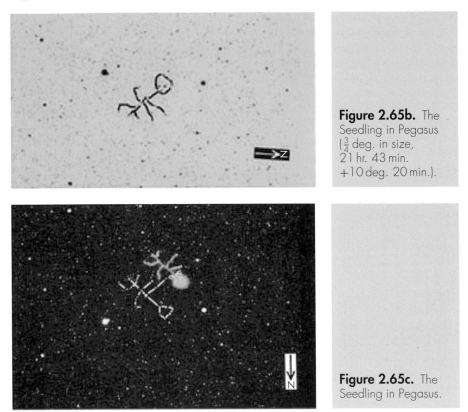

Figure 2.65b. The Seedling in Pegasus ($\frac{3}{4}$ deg. in size, 21 hr. 43 min. +10 deg. 20 min.).

Figure 2.65c. The Seedling in Pegasus.

road a few degrees. Besides a medium size telescope, a medium size imagination is required to see this pattern asterisms shape. Some see a long figure eight while other folks spy a pair of eye glasses.

The Mega Star 5 planetarium program shows The Seedling as 16 stars cylinder shaped at 11 magnitudes. Star 8 Pegasus, Enif, is of interest because it is a variable star and a double star. As a double star, it has a 2.4 magnitude star and an 8.9 magnitude star and they are separated by a little over 2 seconds of an arc. The Seedling pattern asterism is a must at star parties where larger telescopes of low focal length can show its shape clearly. This faint asterism is well worth the time to share with others and it shows up just in time for some very late fall gardening.

Cepheus Map and Asterisms

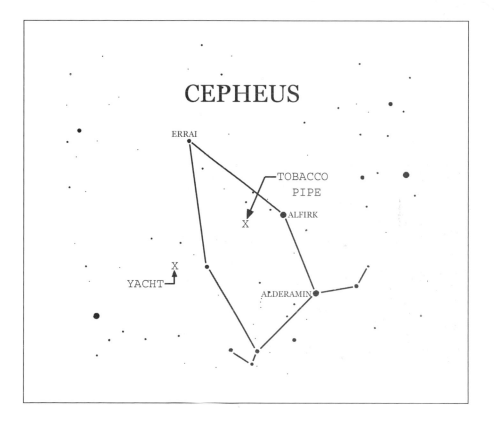

Tobacco Pipe

22 hr. 30 min.   +70 deg. 30 min.

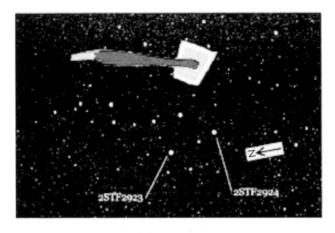

Figure 2.66a.
Double stars.

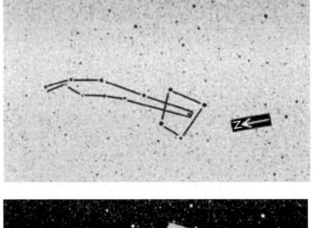

Figure 2.66b.
Tobacco Pipe
Cepheus (1½ deg.
in size, 22 hr.
30 min. +70 deg.
30 min.).

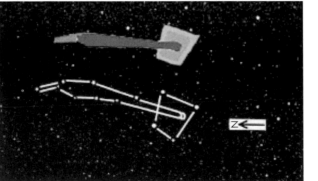

Figure 2.66c.
Tobacco Pipe in
Cepheus.

There is no smoking allowed in this catalog! The cloud of stars surrounding this pipe belongs to the Milky Way. This bright pattern asterism is completely visible in the Sky Atlas 2000 and is located in the smoking section of Cepheus. Seven stars, 8 magnitude and brighter, light up the Tobacco Pipe's, Fig. 2.66, basic shape, while five more stars complete this asterism to 9 magnitudes. An entire can of Prince Albert pipe tobacco would be needed to fill its $1\frac{1}{2}$ degrees of volume. Sit back and relax, viewing is easy with a pair of 50 millimeter binoculars at 10 powers. The Tobacco Pipe is a no brainier visually, located just a few degrees south of star 24 Cepheus and is visible in most any optical finder scope.

The bright stars in this pattern asterism make it a natural for star parties and planetarium programs. There are two bright double stars in this pattern. STF2923 has a 6.3 magnitude star and an 11.3 magnitude star that are separated by $1\frac{1}{2}$ minutes of an arc. STF2924 has a 6.0 magnitude star and a 10.0 magnitude star generously separated by 3 minutes of an arc. Eleven stars are visible that trace out this pattern on the Sky Atlas 2000. So, if you are in the mood but want to quit, this pattern asterism is a habit that will not cause withdrawal.

The Yacht

23 hr. 25 min.   +64 deg. 15 min.

This boat floats just on the border of Cassiopeia and like most yachts, it is big. This 2 degree pattern asterism is on the southeast corner of a mass of stars. The beauty of this

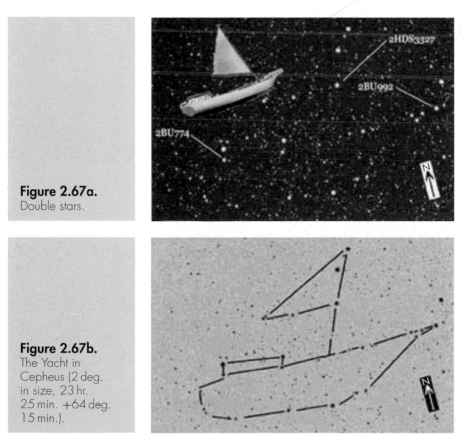

Figure 2.67a.
Double stars.

Figure 2.67b.
The Yacht in Cepheus (2 deg. in size, 23 hr. 25 min. +64 deg. 15 min.).

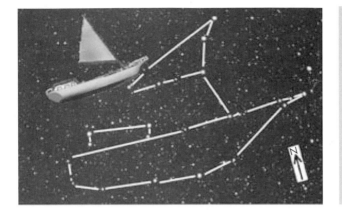

Figure 2.67c. The Yacht in Cepheus.

location is that The Yacht, Fig. 2.67, is sailing under a sky full of stars. The shape starts to appear at 9.5 magnitude with no less then 19 stars which form the sail, cabin, hull and bow. Four more stars complete the shape at one-half magnitudes more. To sea this yacht will require a few episodes of J.A.G. on television and positioning 50 millimeter binoculars at 16 powers on the proper field of stars. It will be smooth sailing if you find M52 in Cassiopeia. Just north of that open cluster is star 4 Cassiopeia. Two degrees north of that star is The Yacht. Use the photo to pick out the sections of this boat and finding it should be a breeze.

The groups of stars that form some of The Yacht are easy enough to find on the Sky Atlas 2000, but the basic shape is too faint to be visible on this atlas. There are four double stars of which I picked out three for reference. They are all very tight doubles. The Yacht pattern will challenge the very best students of dot to dot asterisms, but don't jump overboard on this one. On the Mega Star 5 planetarium program, The Yacht is backed into the corner just above the Cassiopeia border at 10.5 magnitude. If you do have trouble locating The Yacht, then check your position once again. It is very possible that you may have inadvertently sailed into the Bermuda Triangle!

Reference Material

Asterisms at Zenith by Month

Optics Required

JANUARY

Camelopardalis
Vulture — Fifty millimeter binoculars at 16 power.
Perseus
Rabbit — Eighty millimeter binoculars at 20 powers.
Telescopes 4 inch (102 mm) and larger at medium power.

Star Eyed Susan — Eighty millimeter binoculars under 20 powers.
4 to 6 inch (102–152 mm) telescopes at low power.
Eridanus
Letter "F" — Fifty millimeter binoculars at 16 powers.
Letter "D" — Fifty millimeter binoculars at 16 powers.
Any size telescope at low power.

FEBRUARY

Orion
Hand Gun — Fifty millimeter binoculars at 16 powers.
Taurus
Dipper Bowl — Thirty millimeter binoculars at 8 powers.

Fish Hook	Sixty millimeter binoculars at 12 powers.
The Angel	One hundred millimeter binoculars at 20 powers. 5 or 6 inch (127 or 152 mm) wide field refractor at low power.
Poodle	Fifty millimeter binoculars at 16 powers. 4 inch (102 mm) telescope at a low power for a good view.
Barbecue Fork	Eighty millimeter binoculars at 20 powers. 5 inch (127 mm) wide field telescopes are invited.

Lepus

Funnel	Sixty millimeter binoculars at 20 powers. Telescopes of any size when using low power.

MARCH

Lynx

The Turtle	Sixty millimeter binoculars at 16 powers.
Outline Face	Eighty millimeter binoculars at 16 powers or a 4 inch (102 mm) refracting telescope at low power.

Cancer

Fish Hooked	Eighty millimeter binoculars at 16 powers.
Closed Flower	Fifty millimeter binoculars at 10 powers.

Monoceros

Christmas Tree	Fifty millimeter binoculars, 10–20 powers. Telescopes of any size at low power.
The Tooth	Eighty millimeter binoculars at 16 powers. 4 to 6 inch (102–152 mm) telescope at low power.
Loch Ness Monster	At least fifty millimeter binoculars at 10 powers. 3 to 10 inch (76–254 mm) telescope at low power.

Puppis

Emission Comet	Sixty millimeter binoculars at 16 powers.

APRIL

Ursa Major

Spade	Eighty millimeter binoculars at 16 powers.
Gas Pump Handle	Fifty millimeter binoculars at 14 powers.

Sextans

Enter Arrow	Sixty millimeter binoculars at 16 powers. 4 to 8 inch (102–203 mm) telescope at low power.

Hydra

The Moth	Eighty millimeter binoculars at 14 powers. All telescopes are invited.
Umbrella	Fifty millimeter binoculars at 10 powers.
Rocking Horse	Fifty millimeter binoculars at 10 powers.
The Crown/Flower Vase	3 inch (76 mm) wide field refractor at 16 powers.
The Heart	4 to 6 inch (102–152 mm) telescope at low power.
Vacuum Cleaner	Fifty millimeter binoculars at 10 powers.
Sea Horse	Fifty millimeter binoculars at 10 powers.

MAY

Coma Berenices
Curved Arrow | Fifty millimeter binoculars at 10 powers.
Sand Shovel | Fifty millimeter binoculars at 10 powers.
Virgo
Ty's Oil Can | 6 to 10 inch (152–254 mm) reflector at low power.
Phaser Gun | Fifty millimeter binoculars at 10 powers.
Mortar Trowel | Fifty millimeter finder scope or binoculars at 10 powers.

JUNE

Ursa Minor
Engagement Ring | Sixty millimeter binoculars from 14–20 powers. 3 to 6 inch (76–152 mm) wide field refractor at low power.
Soaring Owl | One hundred millimeter binoculars at 20 powers. 5 or 6 inch (127 or 152 mm) wide field refractor at low power.
Shark | Fifty millimeter binoculars at 12 powers.
Draco
Dog and Stick | Sixty millimeter binoculars at 18 powers.

JULY

Hercules
Chinese Kite | Fifty millimeter binoculars at 16 powers.
Candle and Holder | Fifty millimeter finder or binoculars at 10 powers.
Drill Gun | Sixty millimeter binoculars at 16 powers. Medium size telescopes also welcomed at low power.

Ophiuchus
Kermit the Tadpole | Eighty millimeter binoculars at 20 powers. Medium size telescopes at low power a plus.

Scorpius
False Comet | No optical aid, or opera glasses only.
Serpent's Tail | Fifty millimeter binoculars at 10 powers.
Butterfly | Fifty millimeter binoculars at 20 powers. Spotting scopes and telescopes can join in.

SEPTEMBER

Cygnus
Bent Fan | Eighty millimeter binoculars at 16 powers.
Horse's Head | Fifty millimeter binoculars at 16 powers.
Vulpecula
Coathanger | Thirty-to-fifty millimeter binoculars at 8–14 powers.
Aquila
Tennis Racket | Sixty millimeter binoculars at 16 powers. Medium size telescopes at low power.

OCTOBER

Cepheus

Tobacco Pipe	Thirty-to-fifty millimeter binoculars at 8–14 powers.
The Yacht	Fifty millimeter binoculars at 16 powers.

NOVEMBER

Pegasus

Robin Hood Hat	Sixty millimeter binoculars at 14 powers.
	Wide field 4 inch (102 mm) refractor at low power.
The Seedling	6 to 8 inch (152 to 203 mm) telescope at low powers.

DECEMBER

Cassiopeia

E.T./Owl	Sixty millimeter binoculars at 30 powers.
	Any telescope with a pinch of power.
Wagon Wheel	A 50 millimeter finder or binoculars at 10 powers.
Palm Sander	A 50 millimeter binoculars at 16 powers.
Aladdin's Lamp	Fifty millimeter binoculars at 10 powers. 4 inch (102 mm) telescope at low power for good visual effect.
Jelly Fish	Sixty millimeter binoculars at 16 powers.
Measuring Scoop	Thirty-to-fifty millimeter binoculars at 10 powers.
Party Balloon	Fifty millimeter binoculars at 16 powers.
Bowl of Stars	Fifty millimeter binoculars at 10 powers, but 80 millimeter binoculars at 16 powers for open cluster.

Andromeda

Pot and Pitcher	Fifty millimeter binoculars at 12 powers.

Pisces

Spatula	Sixty millimeter binoculars at 16 powers.

Cetus

Hercules Keystone	Fifty millimeter binoculars at 10–20 powers.
Pouring Cup	Sixty millimeter binoculars at 16 powers.

A Glossary of Deep Sky Objects in this Catalog

Name	Object	Figure	Page
E.T./Owl Cluster	NGC 451	2.3	18
Wagon Wheel	NGC 663	2.4	19
Palm Sander	St-5/o.c.	2.5	20
Jelly Fish	St2/o.c.	2.7	23
Party Balloon	M 52	2.9	26
Bowl of Stars	NGC 7789	2.10	27
Rabbit	TR2/o.c.	2.16	45
Star Eyed Susan	M 34	2.17	45
Dipper Bowl	M 45	2.20	52
The Angel	NGC 1647	2.22	54
Vulture	NGC 1502	2.25	60
Hand Gun	CR-69/o.c.	2.26	64
Christmas Tree	NGC 2264	2.28	72
Emission Comet	NGC 2467	2.31	78
The Moth	M 48	2.32	82
Closed Flower	M 44	2.41	97
Ty's Oil Can	M 104	2.47	112
Sand Shovel	M 53	2.51	119
False Comet	NGC 6231	2.52	122
Butterfly	M 6	2.54	124
Kermit the Tadpole	NGC 6633	2.59	138
Coathanger	CR-399	2.60	142
Horse's Head	M 39	2.63	151

A Glossary of Double Stars

Object	Pattern Asterism	Page
A770	Horse's Head	151
A771	Horse's Head	151
A2734	The Tooth	73
AG311	The Angel	54
ARA23	Funnel	68
ARG72	Rocking Horse	85
B1328	Serpent's Tail	123
BAL22	Letter D	49
BLL48	Bent Fan	150
BU111	Enter Arrow	108
BU278	Party Balloon	26
BU464	Kermit the Tadpole	138
BU774	The Yacht	159
BU873	Palm Sander	20
BU992	The Yacht	159
BU1084	Mortar Trowel	115
BU1326	Drill Gun	135
BUP70	The Angel	54
CHR29	Spade	104
CHR130	Closed Flower	97
COU1894	The Turtle	100
ENG61	Dog and Stick	128
ENG72	Bent Fan	150
ES1882	Vulture	60
FOX203	Dog and Stick	128
GCB31	Kermit the Tadpole	138
KU1	Shark	40
H23	E.T. / Owl	18
H109	Fish Hooked	96
HD0309	Serpent's Tail	123
HDS526	Vulture	60
HDS1838	Sea Horse	92
HDS1865	Phaser Gun	113
HDS1881	Mortar Trowel	115
HDS2605	Drill Gun	135
HDS3067	Robin Hood Hat	154
HDS3076	Robin Hood Hat	154
HDS3327	The Yachat	159
HJ110	Palm Sander	20
HJ454	Closed Flower	97
HJ838	Enter Arrow	108
HJ1094	Pot and Pitcher	34

Object	Pattern Asterism	Page
HJ1123	Star Eyed Susan	45
HJ1957	Hercules Keystone	14
HJ2154	Star Eyed Susan	45
HJ2155	Star Eyed Susan	45
HJ2871	Coathanger	142
HU105	Barbecue Fork	57
HU235	Chinese Kite	132
HU1323	Party Balloon	26
HU1500	Sea Horse	92
J73	Fish Hooked	96
J1482	The Tooth	73
J2860	Loch Ness Monster	74
MCA5	Pouring Cup	15
MCA20	Hand Gun	64
MLR383	Aladdin's Lamp	22
OA981	Measuring Scoop	25
RST3603	Umbrella	83
RST3714	Vacuum Cleaner	91
S404	Pot and Pitcher	34
S585	Rocking Horse	85
S798	The Seedling	155
STF8	Dipper Bowl	52
STF9	Fish Hook	53
STF93	Engagement Ring	38
STF138	Spatula	30
STF262	Aladdin's Lamp	22
STF433	Letter F	48
STF641	Fish Hook	53
STF674	Poodle	56
STF680	Poodle	56
STF787	Barbecue Fork	57
STF988	The Tooth	73
STF1260	Umbrella	83
STF1261	Umbrella	83
STF1272	The Turtle	100
STF1308	Crown/Vase	86–88
STF1473	Vacuum Cleaner	91
STF1474	Vacuum Cleaner	91
STF1664	Ty's Oil Can	112
STF1685	Curved Arrow	118
STF1728	Sand Shovel	119
STF1734	Phaser Gun	113
STF2277	Candle and Holder	133
STF2317	Drill Gun	135
STF2521	Coathanger	142

A Glossary of Variable Stars

Object	Pattern Asterism	Page
alf	Engagement Ring	38
BU Tau	Dipper Bowl	52
BM Sco	Butterfly	124
BR Cnc	Closed Flower	97
BV653	The Tooth	73
CD Tau	Poodle	56
DZ Dra	Dog and Stick	128
EP Cnc	Closed Flower	97
eps Peg	The Seedling	155
HD13611	Pouring Cup	15
HD30455	The Angel	54
HD34719	Poodle	56
HD83754	Heart	89
HD170010	Kermitt the Tadpole	138
HR3564	Rocking Horse	85
iot Cas	Aladdin's Lamp	22
LL10492	Hand Gun	64
OP Her	Chinese Kite	132
OR Pup	Emission Comet	78
rho Cas	Bowl of Stars	27
S Mon	Christmas Tree	72
SZ Cam	Vulture	60
UX Lyn	Outline Face	101
V0373	Bowl of Stars	27
V0466 Cas	E. T. / Owl	18
V0480 Tau	The Angel	54
V0605	Jelly Fish	23
V0647	Dipper Bowl	52
V0684	Christmas Tree	72
V0771	Chinese Kite	132
V0724	Fish Hook	53
V0970 Sco	Butterfly	124
V1007	False Comet	122
V1024	Fish Hook	53
V1034	False Comet	122
V1644	Bent Fan	150
V1679	Bent Fan	150
Z1994A	Phaser Gun	113
00466	E. T. / Owl	18
101G Ori	Hand Gun	64
1280	Horse's Head	151
1287	Horse's Head	151
259.1934	Spatula	30
50229	Gas Pump Handle	105

About the Author

At the age of 14 I became aware of the sky above me. The next day I announced, as a student in my seventh grade science class, that I had found constellation coffee pot. A brief burst of laughter ensued. I was politely told to take my seat. The astronomy bug had bitten and I realized, many years later, that there was no cure. Indeed, I still see the Orion asterism as a coffee pot. During my life I would periodically move away from sky glow in an effort to maintain dark skies. I moved from Eastern New Jersey to the southern shores of that state and then to western New Jersey. As years went on this practice became futile and I longed to move to the dark skies of the southwest. Needless to say, the love affair between an observational astronomer and a dark sky is as compelling as trout to their native spawned stream.

Among the many objects observed in the night sky during many years of amateur astronomy, individual preferences eventually flourish and my Achilles heel was the open cluster. I remember vividly displaying my black & white Tri-X ASA 400 collection of open clusters mounted in a slide projector on the dining room wall at home for friends. After one half hour of impressive bright dots, I illuminated the room to find that my audience had fallen asleep; success! Pointing out shapes in groups of stars back then, I realized, needed some serious germination. In 1999 I moved my family to Graham Country, Arizona, just south of Safford. Two reasons brought me to this place; dark skies and Discovery Park. A third reason was added soon after I had joined Discovery Park as a volunteer telescope operator and that was the Mount Graham International Observatories. My fondest wish as a very young amateur astronomer was to stand next to the most powerful telescope in the entire world. At the 10,500 foot elevation, the twin 8.41 meter mirrored Large Binocular Telescope Observatory satisfied that wish. Also, having the honor of being a tour guide to the M.G.I.O. has given me that feeling of where do you go from up?

The Pattern Asterisms project was, for me, simply a glorious accident waiting to happen. Like many observational astronomers, we consider ourselves ambassadors to the professional astronomers. We digest their findings and assimilate this data into people friendly information for telescope events. Some observational astronomers perform good science with their equipment and this data is available for the professional astronomers, but most cannot. Groups of stars in their many shapes, apparent sizes in the sky and magnitudes offer a new world to explore for the non-professional astronomer. This certainly is an outlet into a universe of objects yet to be found. As for me, I now have many new friends in the sky because of my work in stellar shapes.

While searching the sky for star patterns, I realized just how exuberant William Herschel must have felt. Every move of the mount, every new field of view entering the eyepiece of his telescope must have been filled with anticipation. For me, pattern asterisms are a totally new prey in a familiar hunting ground. I imagined having an atlas of known constellations, but void of everything else so that I may plot my new objects into. My guidelines were direct and straightforward. I was seeing a sky of stars in a completely different vantage and that was exciting, to say the least. The professional astronomers may formulate and theorize all they wish and that's a good thing. As for me, all I need is my imagination and a sky full of stars to ponder. It is my sincere hope that this catalog will accelerate an abundance of new patterns, large and small, bright and faint, from every observational astronomer who gazes up at the night sky.

Sincerely, John A. Chiravalle